New Zealand Birds

An identification guide

Stuart Chambers

Published by Reed Books, a division of Reed Publishing (NZ) Ltd,
39 Rawene Road, Birkenhead, Auckland 0626.
Associated companies, branches and representatives throughout the world.

ISBN 13: 978 0 7900 1144 8
ISBN 10: 0 7900 1144 1
First published 2007

A catalogue entry for this book is available from the National Library of New Zealand

Cover design by Cheryl Rowe

Printed in China

Contents

List of birds

Black

Gardens, open country and forests

Wetlands, lakes and rivers

Coast, estuaries and harbours

Sea

Black and chestnut

Black and white

Forests and open country

Coast, estuaries and harbours

Brown

Acknowledgements

For many people birds are the soul of the outdoor environment. A beach without gulls, a mudflat without godwits, a lake without ducks, a lawn without a blackbird or a kowhai tree without a tui is somewhat lessened. Many who are aware of this get an uplifted feeling from birds and wish to learn further. The aim of this book is to do that.

I am very grateful therefore to Peter Dowling and Carolyn Lagahetau at Reed Publishing for taking on this very large project and giving these people an easier approach to discovering birds.

Through my life, from school days, through a farming career to projects and holidays in other countries, I have enjoyed discovering and learning about birds. For me bird-watching has become a major leisure time activity, helped over those years by many knowledgeable people including Sir Charles Fleming, A.H. Hooper, Ross McKenzie and R.B. Sibson. For those not so fortunate to have such mentors I hope this book will fill this gap and be a help and substitute.

In compiling this book I am grateful to my wife Alison and son Richard Chambers for their ideas and help, especially as to bird colours (not an easy task). I am also indebted to Robin Bush, Simon Fordham, Ray Wilson and Ian Southey who provided the majority of pictures. Without their cooperation I could not have produced it. I would also like to thank Dennis Buurman, Bill Jolly, Peter La Tourrette, Ian Montgomery, Ian Tew, Mike Fuller and the Department of Conservation for filling some gaps.

Graeme Leather has provided the imaginative layout and design giving this book its exciting, colourful and attractive appearance. David Orchard provided the map of New Zealand, Brian Parkinson gave continual encouragement and Gillian Vaughan was a great help in finding photos from the Ian Southey collection.

Stuart Chambers
Orewa

Introduction

This book aims to help beginners learn about the birds they see around them. Many people can't get started on bird-watching because they don't know where to commence in terms of identification. It can be difficult to learn about even those birds commonly seen in the home garden or nearby parks if identification is a problem.

Quick guide

The learner is assisted with information in the following ways:

Colour

Pages are colour-edged. The most conspicuous colours of the bird are shown on the edges of the page.

Habitat

This indicates the type of place a particular bird species is most likely to be seen. Most share several habitats; e.g. the Blackbird can be found in gardens, forests and on pasture, or Cook's Petrel feeds from the sea but nests on land. The section heading, with each bird colour, also gives the predominant habitats for birds in that section. For each individual bird species, the habitat that is most frequented is given at the top of the page, but further details are listed under habitat for each bird.

Bird habitats include the following:

- city
- gardens and parks
- native forests
- exotic forests
- ponds, lakes and rivers
- swamps and wetlands
- open countryside
- alpine areas
- scrublands
- sea
- coastal areas
- estuaries
- shoreline.

Range

This indicates how widespread the bird species is and the locations in New Zealand in which the bird may be seen.

Size

Colour sections start with the smallest bird in that section and get progressively larger. This size order is used because most beginners start looking at the smaller

garden birds first and progress later to the larger seabirds. All are compared with the size of a House Sparrow. Measurements are in millimetres from tip of bill to end of tail.

Abundance

This relates to the population status of the species and gives a quick guide as to whether the viewer is looking at a common species or a rare one.

Status

This indicates whether the species is endemic (found only in New Zealand), native (found its own way here from Australia or other countries, where it also lives) or introduced (has been brought here by people).

Target localities

These are brief suggestions at the end of each species offering a locality where a species might be found. Many birds need no such locality guide as they are common everywhere, while others are rare and do need a guide. For example, Red-billed Gulls are found on most beaches and Blackbirds and Chaffinches are common almost everywhere. Kokako and Saddleback, however, can be seen easily only at a place such as Tiritiri Matangi Island.

No.	Name
1	Cape Reinga
2	Aroha Island
3	Russell
4	Cape Brett
5	Rawene
6	Helena Bay
7	Poor Knights Islands
8	Trounson Park
9	Little Barrier Island
10	Sandspit
11	South Head

No.	Name
12	Kawau Island
13	Straka Refuge
14	Wenderholm Park
15	Tiri Tiri Matangi Island
16	Muriwai Beach
17	Kawakawa Bay
18	Pauanui
19	Firth of Thames
20	Miranda
21	Waikato River
22	Meremere

No.	Name
23	Rangiriri
24	Raglan Harbour
25	Maketu
26	Matata
27	Ohope Estuary
28	Kaingaroa Forest
29	Benneydale
30	Opepe
31	Tutira
32	Ohakune
33	Ahuriri Estuary
34	Manawutu Estuary
35	Waikanae Estuary
36	Kapiti Island

New Zealand – South Island

With some locations mentioned in the book

No.	Name	No.	Name
1	Okarito	10	Eglinton Valley
2	Kaikoura	11	Homer Tunnel
3	St Anne's Lagoon	12	Nugget Point
4	Akaroa Harbour	13	Taiaroa Heads
5	Lake Ellesmere	14	Papatowai
6	Otira	15	Bluff
7	Twizel	16	Ulva Island
8	Waikouaiti	17	Waituna Lagoon
9	Haast Pass		

Conspicuous features and characteristics

These give more help in identifying a bird.

Differences between similar species

This information is provided where necessary to help distinguish between species. (Note: to assist users, the common names of species are given with initial capital letters, e.g. Brown Creeper.)

How to use this book

When you see a bird, make notes of the features – main colour, approximate size, any conspicuous feature and habitat. Next, proceed to the colour section and then colour page that most closely matches the bird seen. Check the photograph, habitat and other information to help with your identification.

Bird-watching requirements

The following items make bird-watching easy once the learning process has started:

Guides

Field guides and photo guides provide an easily accessible source of information about the species you see. They include:

- *The Field Guide to the Birds of New Zealand* by Barrie Heather and Hugh Robertson (Viking, 2000) provides descriptions of all birds recorded in New Zealand plus notes about their breeding and habitat details. It can also be purchased in a slim volume, *Hand Guide to the Birds of New Zealand*, which contains bird illustrations and brief descriptions.
- *Birds of New Zealand – Locality Guide* by Stuart Chambers (Arun Books, 2001) provides a total bird-watchers' package in the one volume. It gives detailed descriptions of each bird and points the bird-watcher to localities where the bird may be found.
- *Reed Field Guide to New Zealand Birds* by Geoff Moon (Reed, 1998).
- *Reed Nature Series: Common Birds in New Zealand 1* and *2* plus *Rare Birds of New Zealand* by Geoff Moon (Reed, 1995).
- *Reed Handbook of Common Birds in New Zealand* by F.C. Kinsky and C.J.R. Robertson (Reed, 2000).
- *The Reed Field Guide to Common New Zealand Shore Birds* by David Medway (Reed, 2000).
- *Introducing New Zealand Birds* by Alina Arkins and Len Doel (Reed, 2005).

Habitats

As listed earlier, there is a wide range of bird habitats. Each supports a number of species. For example:

- Estuaries and mudflats have shags, gulls, New Zealand Dotterels, Banded Dotterels, Wrybills, Bar-tailed Godwits, Lesser Knots, Pied Oystercatchers and others. Special birds, which can occasionally be found in such places, include rare migrants such as Golden Plover, Red-necked Stint, Sharp-tailed Sandpiper and Curlew Sandpiper. Because of this variety, these areas can provide a challenge for bird-watchers.
- Forests, including pine forests, have native species such as New Zealand Pigeon, Tui, Bellbird, Tomtit, Robin and Whitehead. Sightings of Kaka, Kakariki, Long-tailed Cuckoo, Rifleman and New Zealand Falcon are less common in pine forests.
- Gardens and city parks have mainly introduced birds such as the various finches, Blackbird, Song Thrush, Starling and House Sparrow. However, many gardens also have native species such as New Zealand Pigeon, Morepork, Grey Warbler, Tui, Bellbird, Silvereye and Kingfisher.
- Lakes and rivers commonly have the introduced Mallard and Black Swan. However, native species such as Grey Duck, Shoveler, Grey Teal, Scaup and Dabchick are what bird-watchers look for here.
- Oceans around New Zealand, with their scattering of offshore breeding islands, make New Zealand a rich locality for seabirds. Birds feed on the oceans and nest or roost on the islands. Common shoreline species here are gulls, gannets and shags but in deeper waters these soon give way to albatrosses, petrels, shearwaters and prions. The true seabirds are the birds that bird-watchers seek.
- Open country and pasturelands mostly have introduced varieties, such as the Skylark, Blackbird, Song Thrush, various finches, Starling, Magpie and Rook, with Mallard present anywhere where there is water. The main native species are the Australasian Harrier, with Pukeko on pasture verges. Spur-winged Plover, Pied Stilt and Pied Oystercatcher also frequent pasture areas in some localities, mostly in winter.
- Scrublands, areas of manuka and low-growing vegetation are usually home to the Grey Warbler. In these areas bird-watchers also look for the Pipit and Fernbird.
- Swamps and wetlands commonly have Pukeko, but bird-watchers here look for the Banded Rail, Fernbird, Spotless Crake, Marsh Crake and Bittern.

Always have an open mind when bird-watching. A common Blackbird could show up in a wetland, a Kaka in an urban garden, a Spotted Shag or a Caspian Tern at an inland lake, or a Black Swan or Shoveler on a stretch of coast. Anything is possible, but take careful notes so later you can verify your sightings.

Binoculars and telescopes

For finding birds in the bush, 8 x 40 binoculars are sufficient. These have a wide field of vision which will help the beginner to find the bird more easily. They also let in plenty of light, which makes the bird look sharper and allows sharper focus. When bird-watching over lakes or estuaries, a telescope is essential for accurate observation. A 25x magnification is usually sufficient.

Notebooks

Always carry a notebook with you when bird-watching and make notes as you go. Notes should include habitat, conspicuous features and characteristics, date, place, weather conditions and the state of the tide when viewing over estuaries. Small spiral-bound notebooks with a pencil attached are best.

Final entries made later could include:

- a general note with date
- a species note (e.g. Tui)
- a locality note (e.g. Miranda).

This recording in triplicate allows for fast retrieval of facts at a later date.

Starting birding

Preparation for an excursion to a birding spot

There are two ways to embark on a bird-watching adventure:

- Select a locality and visit it to see what can be found.
- Select a bird species and go to a locality which has this bird and try to find it.

For beginners, the best way is to select a locality and go and see what can be found. As the bird list grows, localities can then be selected with a particular species in mind.

The first locality to visit

Beginners should look in the gardens around their homes first and see what is feeding on the lawn. Usually most lawns have House Sparrows, Starlings, Blackbirds and Song Thrushes. In winter they also have Chaffinches, Goldfinches, Greenfinches and Silvereyes. All of these can present identification problems. For example:

- distinguishing the brownish female Blackbird from the Song Thrush, especially during the moulting season
- distinguishing the juvenile greyish Starlings from their black-coloured parents
- distinguishing the brownish female House Sparrow from a Redpoll or a Hedge Sparrow.

This shows there is plenty to do right alongside the house. When things are sorted out there, it is then time to branch out and try a locality elsewhere.

Fantail (black phase)

Conspicuous colour Black.
Habitat Suburban gardens, scrublands, wetlands, native and exotic forests.
Range Throughout New Zealand but mainly in the South Island. Absent from alpine areas and areas of vast grassland.
Size 160 mm (House Sparrow 145 mm, Grey Warbler 100 mm).
Abundance Seldom seen but not uncommon.
Status Native – a black phase of the Fantail.

Family Monarchidae

Species *Rhipidura fuliginosa fuliginosa*

Common names Black Fantail, South Island Fantail. (There are several common Maori names, varying from one part of the country to another, including Piwakawaka, Piwaiwaka, Tiwakawaka, Tiwaiwaka, Tirairaka.)

Phases The Fantail has a pied phase and a black phase. The black phase is more common among South Island birds, especially in areas around the Marlborough Sounds, south of Christchurch and north of Dunedin. In the North Island, black birds are rarely encountered.

Subspecies Three subspecies are recognised: North Island Fantail (*R. f. placabilis*), South Island Fantail (*R. f. fuliginosa*), Chatham Island Fantail (*R. f. penitus*).

Description Upperparts and underparts Black. Usually a faint white spot can be seen behind the eye.

Call A high-pitched 'cheet cheet' communication call. When in nesting territories male birds have a vocal and constant repetitive chattering 'tweetatweetatweet' song call.

Nest Neat wine-glass-shaped cup of twigs and leaves lined with moss. Usually four white, brown-speckled eggs are laid.

Target localities Botanic Garden, Wellington ✎ Forested areas from Picton to Havelock in the Marlborough area of the South Island.

Starling (breeding)

Conspicuous colour Black.
Habitat Gardens, pasture, open areas.
Range Found throughout New Zealand away from forests and alpine areas.
Size 210 mm (House Sparrow 145 mm).
Abundance Common.
Status Introduced.

Family Sturnidae

Species *Sturnus vulgaris*

Common name Starling

Description – breeding
Upperparts and underparts Glossy black with a purple sheen. There is some faint remnant whitish spotting on flanks. Wings Black, with a greenish sheen and feathers edged with brown. Bill Yellow.

Description – winter bird
Upperparts Black, with buff-coloured head and white spotting to the neck. Wings Black, feathers edged with light brown. Underparts Blackish with very conspicuous white spotting. Note that the female is more spotted than the male bird. Bill Brown.

Description – immature Upperparts Greyish-brown, with some darker wing striping. Underparts Greyish. Eye Black, with a dark-brown surround. Bill Brown.

Conspicuous features Yellow bill in the breeding season ✎ White speckling on birds not in full breeding plumage ✎ Dark eye on juvenile birds.

Conspicuous characteristics Outside of the breeding season Starlings flock by night in large communal roosts, gathering in groups before dusk ✎ Birds feed in paddocks in large groups outside of the breeding season (If disturbed, these flocks look spectacular as they take to the air) ✎ In the breeding season birds constantly fly backwards and forwards to nests.

Call The Starling is a recognised mimic of such birds as the California Quail, Shining Cuckoo, and the alarm calls of the Blackbird. Its own song is a warble of a variety of thin notes uttered from a high perch. Wings are clapped when the bird is singing.

Nest A bundle of straw and sticks in holes or under the eaves of buildings. Egg laying starts in the first week in October. Up to six pale-blue eggs are laid.

Target localities Found everywhere from towns to extensive pasturelands.

Blackbird (male)

Conspicuous colour Black.
Habitat Gardens, pasture, native and exotic forests.
Range Found throughout New Zealand and on offshore islands.
Size 250 mm (House Sparrow 145 mm, Song Thrush 230 mm).
Abundance Common.
Status Introduced.

Family Muscicapidae

Species *Turdus merula*

Common name Blackbird

Description – male
Upperparts and underparts Black.
Eye-ring Yellow.
Bill Bright yellow in the breeding season.

Description – female
Upperparts Dark brown.
Chin Grey.
Breast and underparts Light brown, speckled with dark brown.
Bill Orange in breeding season, otherwise brown.

Description – immature Upper and underparts Dark brown with some breast speckling.

Conspicuous features Yellow eye-ring and bill.

Conspicuous characteristics Commonly feeds on lawns and pasture ✎ When on the ground it hops rather than walks ✎ Birds pause when feeding, often with head turned and ear to ground ✎ When disturbed, birds will fly off, making an alarm call ✎ Have a habit of sunbathing, lying under full sun with wings spread ✎ Birds moult heavily over late summer, more so than the Song Thrush.

Call Song period is from August through to January but the 'tok tok tok tok' alarm call can be heard at any time of year. The song of the male Blackbird is melodious and fluid with notes uttered in phrases with distinct pauses between each. It does not repeat notes as does the Song Thrush. Instead it runs a variety of notes together. Blackbirds sing most strongly at dawn and dusk.

Nest A bulky cup of grass, twigs and leaves bound with some mud and lined with fine grasses. Up to four greenish, brown-speckled eggs are laid. Both male and female birds help with nest building and the feeding of young.

Target localities Found almost everywhere but not above the snow line.

Tui

Conspicuous colour Black.
Habitat Gardens, parks, forests.
Range Well-spread throughout the North Island, although absent in some intensively farmed areas. In the South Island it is found throughout most forested areas but is absent from Canterbury and is seldom seen in the beech forests of Fiordland. It is common on Stewart Island.
Size 300 mm (House Sparrow 145 mm, Blackbird 250 mm).
Abundance Common.
Status Endemic.

Family Meliphagidae

Species *Prosthemadera novaeseelandiae*

Common names Tui, Koko (Parson Bird is an old settlers' name, now rarely used.)

Description Upperparts and underparts Black, with green-purple iridescence. Wings Black, with white wingbars, conspicuous when in flight. Neck and nape Black, with a lacy white collar. Throat Black, with two white throat tufts. Bill Black.

Conspicuous features A bird of glossy plumage with considerable iridescence in good sunlight — Has conspicuous white-feather throat tufts — Slightly larger than the common Blackbird — White wingbars visible in flight.

Conspicuous characteristics Flight is jerky and erratic — After about six wing-beats, the bird pauses in flight before resuming — Will rise vertically to great heights, with very rapid wing-beats, before plunging down again — Noisy flier and especially fast and manoeuvrable when chasing unwanted birds from nesting territories — Regularly feeds on nectar from flax (*Phormium tenax*), five-finger (*Pseudopanax* species), kowhai (*Sophora* species) and the coastal pohutukawa tree (*Metrosideros excelsa*).

Call The song, delivered by both sexes, is of bell-like notes, similar to the Bellbird, but less pure, incorporating many guttural notes and gurgles. Some notes are beyond human hearing. Dialects have developed in some areas. When in full song, male birds sing from vantage points. The song is constant, regular, guttural and repetitive. The female usually utters a melodious phrase of up to eight notes and then is silent. The male Tui is one of the first birds to start singing in the morning, uttering its first notes at least an hour before sunrise. In summer, birds will sing from daybreak to dusk. Birds sing throughout the year.

Nest A platform of sticks, lined with fine grasses, in a fork of a tree or outer branch at about 4 m above ground. Up to four white, brown-blotched eggs are laid.

Target localities Wenderholm Regional Park north of Auckland city — Pelorus Bridge, Marlborough.

Rook

Conspicuous colour Black.
Habitat Open country, not venturing into towns and cities as do Australian Magpies. Also coastal areas and sandy beaches.
Range Hawke's Bay province to southern Hawke's Bay and Woodville. Scattered colonies are found in the Manawatu, Matamata, Hauraki Plains and Miranda areas. In the South Island mainly in mid-Canterbury and on Banks Peninsula.
Size 450 mm (House Sparrow 145 mm, Australian Magpie 410 mm).
Abundance Uncommon due to control.
Status Introduced.

Family Corvidae Species *Corvus frugilegus* Common name Rook

Description Face An area of rough skin on the face beneath the eye is greyish. Upper and underparts Glossy black. Wing Black and 'fingered' at tips. Legs Greyish-brown with shaggy black-feathered thighs. Bill Greyish with dark tip. Tail Black and rounded at tip.

Conspicuous features A black bird ✎ Greyish face and bill.

Conspicuous characteristics Birds feed on the ground, usually in groups but occasionally alone ✎ Feeding birds often play, chasing each other into the air and tussling ✎ At dusk birds seek communal roosts.

Call A 'caw caw' is the usual call. Birds are extremely noisy when they leave their communal rookeries at daybreak.

Nest Birds are communal nesters. Nests are made of grass and twigs and placed high up in trees, usually conifers. Up to four bluish, brown-blotched eggs.

Note Rooks were introduced from England in the 1860s. They established colonies in both the Hawke's Bay and Mid-Canterbury districts. Because they were blamed for damage to arable crops, many thousands were poisoned and numbers restricted.

Target locality Miranda, Firth of Thames.

Wild Turkey

Conspicuous colour Black.
Habitat Farmyards and open country.
Range Selected areas in North Island and Marlborough and North Canterbury.
Size 1200 mm (House Sparrow 145 mm, Pheasant 800 mm).
Abundance Common in selected areas.
Status Introduced.

Family Phasianidae **Species** *Meleagris gallopavo*

Common names Turkey, Wild Turkey

Description – male Head and neck Blue and red naked skin. Upperparts and underparts Black, faintly barred diagonally with white. Breast Loose black feathers hang like a beard.

Description – female Face Blue naked skin. Upperparts and underparts Black but paler than male bird.

Conspicuous features Naked head and neck on male bird ✎ Beard of feathers on breast.

Conspicuous characteristics Unusual 'gobble gobble gobble' sound from male bird.

Call 'Gobble gobble gobble' sound from male bird.

Nest A bowl in grasses or low vegetation. Up to 15 cream, brown-blotched eggs are laid.

Target localities Open countryside in Northland.

Spotless Crake

Conspicuous colour Black.
Habitat Wetlands and swamps on the mainland. Inhabits the forest floor on offshore islands such as Tiritiri Matangi and Poor Knights.
Range Throughout the North Island. In the South Island, found in Marlborough, Westland, North Canterbury and Southland.
Size 200 mm (House Sparrow 145 mm, Marsh Crake 180 mm).
Abundance Secretive but not uncommon.
Status Native.

Family Rallidae **Species** *Porzana tabuensis*

Common names Spotless Crake, Puweto

Description Head and nape Black. Back and wings Dull brown. Underparts Dark grey. Eye Red.

Conspicuous features Red eye ✎ Dull brown wing colouring ✎ The brown of back and wings seen in good light only.

Conspicuous characteristics Birds will fly up from cover if startled ✎ Birds are shy in the open and quickly run for cover.

Call 'Mook mook mook mook' often followed by a 'brururu' bubbly-type call. Also isolated 'mook' sounds.

Nest On the ground in grass or rushes. Up to four pinkish-cream eggs are laid.

Target localities Tiritiri Matangi Island, off Whangaparaoa Peninsula near Auckland ✎ Nga Manu Sanctuary, Ngarara Road, Waikanae, north of Wellington ✎ St Anne's Lagoon, Cheviot, North Canterbury.

Australian Coot

Conspicuous colour Black.
Habitat Clean-water lakes or slow-moving rivers where it feeds among the rush and raupo verge, diving for food that is largely of vegetable matter.
Range In the North Island from North Auckland south, and on the eastern side of the South Island from Kaikoura to Oamaru.
Size 380 mm (House Sparrow 145 mm, Blackbird 250 mm).
Abundance Common in specific areas.
Status Native.

Family Rallidae **Species** *Fulica atra* **Common name** Coot

Description Upperparts Black. Underparts Dull black, sometimes tending to greyish. Bill and frontal shield White. Legs and feet Grey with lobes on toes.

Conspicuous features White bill and frontal shield ✎ Lobed toes.

Conspicuous characteristics Flight is low and fast but seldom observed ✎ Usually observed swimming ✎ Has a habit of bobbing its head while it is swimming ✎ Birds regularly dive ✎ Birds bring up food for young and feed them on the water ✎ Birds usually prefer still water and not river currents.

Call A 'krark krark' sound.

Nest A floating platform of rushes and twigs anchored to a rush bush. Up to eight creamy eggs are laid.

Target localities Western Springs Lake, Auckland ✎ Lake Rotoroa, Hamilton ✎ St Anne's Lagoon, Cheviot, North Canterbury.

New Zealand Scaup

Conspicuous colour Black.
Habitat Deep, clean, freshwater lakes throughout New Zealand, including those bordering the coastal sand dunes.
Range Found through many parts of New Zealand. Common in the Rotorua and Taupo areas. Well spread on the lakes of the South Island.
Size 400 mm (House Sparrow 145 mm, Mallard 580 mm).
Abundance Common.
Status Endemic.

Male

Female

Family Anatidae

Species
Aythya novaeseelandiae

Common names
Scaup, Black Teal, Papango

Description — male
Upperparts and underparts Glossy black.
Wings Black, with a broad white wingbar on primary and secondary feathers.
Eye Iris yellow.
Bil Blue-black.

Description — female
Head and neck Dull brown.
Wings and back Black with white wingbar as for the male. Underparts Dull brown. Eye Brown. Bill Black, with a white crescent marking at base of bill.

Conspicuous features White band on wings in flight / Yellow iris on male birds.

Conspicuous characteristics Flight is fast and just above the water / When not swimming together in breeding pairs, birds are usually found in large loafing flocks in sheltered water / When feeding, birds are continually diving and reappearing at about 15-second intervals / Young birds dive along with their parents.

Call When sitting loafing in rafts, male birds continually utter a high-pitched, rattled whistle.

Nest A bowl of grass or rushes, lined with down, and close to the water. Up to eight cream eggs are laid.

Target localities Lake Rotorua, at Rotorua / Lake Wakatipu at Queenstown.

Paradise Shelduck (male)

Conspicuous colour Black.
Habitat Pasture, ponds, shallow freshwater areas and lakes.
Range Throughout New Zealand.
Size 630 mm (House Sparrow 145 mm, Mallard 580 mm).
Abundance Common.
Status Endemic.

Family Anatidae

Species *Tadorna variegata*

Common names Paradise Shelduck, Paradise Duck or Parry, Putangitangi

Description – male Head and neck Black, with a metallic greenish sheen. Upper and underparts Black, lightly barred with white. Wings Black, with prominent white wing-coverts on upper and underwings and a large green speculum. Abdomen Reddish-brown. Undertail Orange-chestnut. Bill Black. Legs and feet Black.

Description – female Head White with black eye. Breast and underparts Orange-chestnut, tending to brownish when not in breeding plumage. Wings Black with prominent white wing-coverts and a large green speculum. Undertail Orange-chestnut. Bill and legs Black.

Description – ducklings Upper and underparts Zebra-striped brown and white when first born.

Description – immature birds Upperparts and underparts Fledglings of both sexes resemble the male.

Conspicuous features White on the head of the female White on the wings of in-flight birds.

Conspicuous characteristics For much of the year this species is usually seen in pairs After the breeding season family parties are encountered When disturbed, birds usually take to the wing and circle while calling in a duet manner Generally, a nervous bird ever alert for people intruding Ducklings have a habit of sitting in little pyramid-like heaps when very young, always within sight of the parent birds. If disturbed they scatter quickly.

Call Male – a deep 'klonk klonk'. Female – a high-pitched 'ziz zik'.

Nest In holes in the ground or in old rotting stumps or rock crevices but sometimes in holes in trees. Up to nine white eggs are laid.

Target localities Lake Rotoehu near Rotorua in summer Matata Lagoons, Bay of Plenty St Anne's Lagoon, Cheviot, North Canterbury Open countryside.

Black Swan (mature)

Conspicuous colour Black.
Habitat Inland lakes, tidal harbours and estuaries.
Range Throughout New Zealand.
Size 1200 mm (House Sparrow 145 mm, Mute Swan 1500mm).
Abundance Common.
Status Introduced.

Family Anatidae

Species *Cygnus atratus*

Common names Black Swan

Description – mature

Upperparts and underparts Black.

Wings Black, with white on primary and some of the secondary feathers.

Bill Red, with a white band near the tip and a white tip.

Legs and feet Black.

Description – immature

Upperparts and underparts Light brown.

Conspicuous features

A large black swan ✎ Has a paint-like white band across the upper mandible ✎ White wing primaries.

Conspicuous characteristics When on the water birds continually raise their wings and flap them ✎ In flight, neck is outstretched ✎ Song and wing whistle are regularly heard overhead at night.

Call The Black Swan has a very musical song usually heard when birds are flying. Also, a musical whistle is uttered when birds are sitting on the water.

Nest A mound of grass and rush stems near the shoreline and on the ground. Up to six pale-green eggs are laid.

Target localities Lake Pupuke, Takapuna, Auckland ✎ Lake Whangape, Rangiriri, North Waikato ✎ St Anne's Lagoon, Cheviot, North Canterbury.

Black Stilt

Conspicuous colour Black.
Habitat Braided South Island rivers in summer. North Island harbours in winter.
Range Mid-Canterbury in summer. Kawhia Harbour in winter.
Size 400 mm (House Sparrow 145 mm, Pied Stilt 380 mm).
Abundance Rare.
Status Endemic.

Family Recurvirostridae **Species** *Himantopus novaezelandiae*

Common names Black Stilt, Kaki

Description – mature Upperparts and underparts Black. Legs Red. Bill Black.

Description – immature Upperparts White. Wings Black. Throat and breast White. Abdomen Black.

Note Pierce (1984) observed the progression of young Black Stilts from the nestling stage to the third summer and recorded a succession of steps by which the young birds progressed from being black-winged only, to black-flanked, to smudgy-breasted, to flecked with grey and finally to becoming totally black.

Conspicuous feature Black colouring is distinctive.

Call A yapping, 'yep yep yep' sound, similar to the Pied Stilt.

Nest A solitary breeder which makes a nest of grass and soft twigs, usually on a riverbed island. Up to four greenish, brown-blotched, eggs are laid.

Target localities Hauturu Road, Oparau, where it touches the Kawhia Harbour (winter only) Ohau River about 5 km south of Twizel (summer only).

Variable Oystercatcher (black phase)

Conspicuous colour Black.
Habitat A bird of coastal areas only, inhabiting sandy beaches, mudflats and rocky promontories.
Range Found around Northland, the Bay of Plenty and down the east coast to Wellington and up to Wanganui, but with a gap around the Taranaki coast. Around the South Island, gaps are found along the coasts of Canterbury and South Westland.
Size 480 mm (House Sparrow 145 mm, South Island Pied Oystercatcher 460 mm).
Abundance Regularly seen in selected areas.
Status Endemic.

Family
Haematopodidae

Species
Haematopus unicolor

Common names
Variable Oystercatcher, Black Oystercatcher (South Island), Torea, Torea-pango

Description – all black phase Upperparts and underparts Black. Bill, legs and eyes Red.

Description – pied phase Upperparts Black. Underparts Variable white but white shoulder tabs do not extend up neck as with the Pied Oystercatcher. Bill, legs and eyes Red.

Conspicuous features Birds look heavier than South Island Pied Oystercatcher ✎ No white shoulder tabs ✎ Red bill, legs and eyes.

Conspicuous characteristics Birds carry on frequent aggressive displays towards each other ✎ Tend to cluster in groups at end of beaches ✎ Will carry out dropped-wing displays to distract intruders from nest.

Call A noisy bird with a call similar to South Island Pied Oystercatcher – a melodic but quite aggressive 'keeleep keeleep' sound.

Nest A scrape on a coastal area with up to three brown, blotched dark-brown eggs.

Target localities Waiwera Spit, Waiwera, North Auckland ✎ Riverton Beach, Southland.

Little Shag (immature black phase)

Conspicuous colour Black.
Habitat Coastlines, estuaries, harbours, inland rivers and lakes.
Range Well-spread throughout New Zealand, including Stewart Island.
Size 560 mm (House Sparrow 145 mm, Pied Shag 810 mm).
Abundance Common.
Status Native.

Family Phalacrocoracidae

Species *Phalacrocorax melanoleucos*

Common names Little Shag, Kawau-paka (also, less common, Little Pied Shag or White-throated Shag)

Phases Four: white-faced; white-faced plus white-throat; white-faced and totally white underparts; white-faced and smudgy underparts. Fledglings come in two phases: black, but separated from the Little Black Shag by a shorter yellow bill (Little Blacks have lead-coloured bills); smudgy white face and all-white underparts with yellow bill.

Description (immature) – black phase Upper and underparts Black.

Description (immature) – smudgy phase Face Smudgy white. Upperparts Black. Underparts White. Bill Yellow.

Description – mature Head White face and throat is common on all phases. The white colouring extends up the face, around the eye to the crown with only a narrow belt of black being visible on the top of the head. A very small crest separates black-phase juvenile birds from Little Black Shags. Upperparts Black. Underparts (See various phases above.) Bill Yellow. Feet and legs Black.

Conspicuous characteristics Coastal as well as inland, unlike Pied Shag, which is only coastal ✎ Sits in social groupings on harbour and lake jetties and on lake shorelines, often with all phases including fledglings in the one group ✎ Nests in loose colonies, sometimes in large numbers ✎ Egg laying is spread from August to March ✎ Birds follow flight paths from breeding colonies to feeding grounds ✎ Birds tend to feed singly, not in packs as the Little Black Shag does ✎ Birds can spend up to 20 seconds under water in each dive for food.

Note Little Shags are widely spread from East Borneo, Java, New Guinea, Australia and New Caledonia to New Zealand. The Australian and the New Caledonian birds are totally pied, having all white underparts.

Call Guttural croaks and squeals at the nesting site only.

Nest A colonial nester, making a twig nest in trees. Up to four bluish-green eggs.

Target localities Orakei Basin, Auckland ✎ Lake Taupo, lakefront ✎ Picton, Marlborough Sounds ✎ St Anne's Lagoon, Cheviot, North Canterbury ✎ Ulva Island, Stewart Island.

Little Black Shag

Conspicuous colour Black.
Habitat Found on harbours and estuaries, inland lakes and rivers, and often near human habitation.
Range Well-spread through Northland, the Waikato and the Bay of Plenty, but scattered populations exist in many other places in the North Island. Away from the Marlborough Sounds it is generally uncommon in the South Island.
Size 610 mm (House Sparrow 145 mm, Black Shag 880 mm).
Abundance Common in the North Island.
Status Native.

Family
Phalacrocoracidae

Species
Phalacrocorax sulcirostris

Common names
Little Black Shag, Kawau-tui

Description — mature
Upperparts Black.
Underparts Black.
Bill Lead colour.
Feet and legs Black.

Description — immature bird
Upperparts and underparts Black.

Conspicuous features
Smaller than the Black Shag ✎ Of sleeker appearance than the Little Shag ✎ Bill is long, slender and lead-coloured ✎ Tail is slightly shorter than that of the Little Shag.

Conspicuous characteristics Often seen swimming in packs ✎ Up to 100 birds might be seen bobbing up together and diving together ✎ A gregarious species which often roosts together in large numbers.

Call Croaks and whistles at the nest site only.

Nest A colonial nester making a twig nest in a tree. Up to four greenish eggs.

Target localities Orakei Basin, Auckland ✎ Waikato River, Hamilton ✎ Lake Rotorua near the Government Gardens, Rotorua ✎ Picton, Marlborough Sounds.

Stewart Island Shag (bronze phase)

Conspicuous colour Black.
Habitat Coastal, favouring rocky islets.
Range The South Island coast from Timaru to Stewart Island and Fiordland.
Size 680 mm (House Sparrow 145 mm, King Shag 760 mm).
Abundance Common locally.
Status Endemic.

Family Phalacrocoracidae

Species *Leucocarbo chalconotus*

Common names Stewart Island Shag, Bronze Shag, Kawau-mapua

Phases Two distinct phases: a pied phase and an all-black phase known as the bronze phase. There are also intermediate phases between the two. The name Bronze Shag was given to the black phase when it was first considered to be a separate species. The name was derived from the green iridescent sheen which covers the black upper feathers of this bird. Bronze, pied and intermediate phases can always be found together in larger colonies such as at Taiaroa Head and on Whero Island off Stewart Island.

Description – bronze phase Head Black. Eye-ring Blue. Caruncles Orange-yellow, tending more to orange than the King Shag. Facial skin Purple. Upperparts Black. Feathering has a noticeable sheen when in breeding plumage. Underparts Black, with an iridescent green sheen to the feathering. Bill Grey. Feet and legs Pink.

Description – intermediate phases Upperparts Black. Underparts White, in varying amounts.

Description – pied phase Head Black. Eye-ring Blue. Caruncles Orange-yellow, tending more to orange than the King Shag. Facial skin Purple. Upperparts Black. Wings Black, with a bold slash of white on scapular feathers. Underparts White from the throat down. Bill Grey. Feet and legs Pink.

Conspicuous features All black ✎ Blue eye-ring.

Conspicuous characteristics Often seen in groups on offshore rocks ✎ Usually seen feeding in deeper water than Pied Shags ✎ A more confiding bird than the King Shag of the Marlborough Sounds.

Call Silent, except when displaying at the colonies, when grunts are made.

Nest On the ground on rock ledges or among rock crevices where it builds a nest made of plant and vegetable material. Two pale-blue eggs are laid.

Target localities Taiaroa Head: birds can be seen from the Royal Albatross observatory. Here both Stewart Island Shags and Spotted Shags can be seen together ✎ Riverton: common here and at Colac Bay. Birds sometimes roost on the beach.

Black Shag

Conspicuous colour Black.
Habitat Muddy estuaries, tidal areas, and sand banks, but also found inland along drains, canals and lakes in most freshwater localities.
Range Found throughout New Zealand.
Size 880 mm (House Sparrow 145 mm).
Abundance Common.
Status Native.

Family Phalacrocoracidae **Species** *Phalacrocorax carbo*

Common names Black Shag, Kawau (Great Cormorant outside of New Zealand)

Description — mature Face White (on some birds only). Facial skin Yellow. Upperparts and underparts Black. Flank patches White (on some birds only). Bill Grey.

Description — immature Upperparts Brownish-black. Underparts Variable amounts of dirty white.

Conspicuous feature Heavy-looking, all-black bird.

Conspicuous characteristics Often seen perched high up on trees, with wings extended ✎ Often seen sitting on sand banks in an upright position with a kinked neck bulge ✎ Heavy and slow flight.

Call Usually silent, but croaks and grunts at the nest.

Nest A colonial nester that makes a platform of sticks in a tree. Up to four bluish-green eggs are laid.

Target localities Miranda, Firth of Thames ✎ Fortrose Estuary, Southland.

Grey-faced Petrel

Conspicuous colour Black.
Habitat Sea.
New Zealand range Common in northern New Zealand waters.
Size 410 mm (House Sparrow 145 mm).
Abundance Common locally at sea.
Status Endemic.

Family Procellariidae

Species
Pterodroma macroptera

Common names
Grey-faced Petrel, Oi, Great-winged Petrel (Australia)

Description
Face and forehead Grey.
Upperparts and underparts Blackish-brown.
Bill and feet Black.

Conspicuous features An all-black bird ✎ Wings are long and narrow ✎ Short, dark bill ✎ Grey face.

Conspicuous characteristics Tosses into the air at a fast pace before plunging again down near the wave crest.

Note In the Hauraki Gulf all four species of these black-coloured petrels and shearwaters can be seen together.

Call A guttural 'oi oi oi' is made by birds when near their burrows.

Breeding islands close to New Zealand Breeds on many offshore islands from Gisborne northwards on the east coast of the North Island to the Three Kings Islands. Breeds from Taranaki north on the west coast. Major breeding islands are the Hen and Chicken Islands and Mokohinau Islands off Northland, Mercury Islands off Coromandel and The Aldermen Islands, Whale Island and White Island off the Bay of Plenty.

Breeding months May to December. Birds return to their breeding grounds in late February. Birds lay one white egg in a burrow.

Range worldwide Ranges in the non-breeding months from New Zealand to South Africa northward to about latitude 20°.

Target locality Seas out from Sandspit, Northland.

Differences between Grey-faced Petrels, Black Petrels, Sooty Shearwaters and Flesh-footed Shearwaters

- Grey-faced Petrels have short, stubby bills when compared with the long, thin bills of the shearwaters.
- Grey-faced Petrels have black bills; Black Petrels have yellowish bills tipped with black; Sooty Shearwaters have dark-grey bills; Flesh-footed Shearwaters have yellowish bills tipped with brown.
- Grey-faced Petrels, Black Petrels and Sooty Shearwaters have black legs and feet which separate them from the Flesh-footed Shearwater, which has flesh-coloured feet.

Flesh-footed

- Grey-faced Petrels have grey faces, noticeable in good light.
- Grey-faced Petrels are fast fliers, tending to bounce off waves to a much higher altitude than Sooty and Flesh-footed Shearwaters, which are more wave skimmers.

- Grey-faced Petrels lack the silver underwing of Sooty Shearwaters.

Short-tailed Shearwater

Conspicuous colour Black.
Habitat Sea.
New Zealand range Northern North Island waters from May to October.
Size 420 mm (House Sparrow 145 mm, Flesh-footed Shearwater 450 mm).
Abundance Common migrant in waters to northeast of the North Island.
Status Native.

Family Procellariidae

Species *Puffinus tenuirostris*

Common names
Short-tailed Shearwater, Tasmanian Muttonbird

Description
Upperparts and underparts Black. Underwing Dark grey (not silver like the Sooty Shearwater). Bill Dark grey (shorter than the Sooty Shearwater). Legs and feet Lilac with brown edges.

Conspicuous features Smaller than the Sooty Shearwater / Lacks silver underwing which separates it from the Sooty Shearwater / Lilac-coloured feet separate it from the brown feet of the Sooty Shearwater / Dark-grey, short bill separates it from the Sooty Shearwater.

Conspicuous characteristics A fast flier similar to the Sooty Shearwater / Stiff-winged flight / Tends to flap and then glide for some distance along wave-tops.

Breeding islands Breeds on the islands of Bass Strait, Australia.

Breeding months November to May. One white egg is laid in a burrow, the first birds arriving back to the breeding colonies in late September.

Range worldwide Migrates into the Pacific Ocean.

Target locality Seas out from Sandspit in Northland from February to May.

Sooty Shearwater

Conspicuous colour Black.
Habitat Sea.
New Zealand range Widely spread around all New Zealand waters. At times can be seen feeding close to land, especially when weather is rough.
Size 440 mm (House Sparrow 145 mm, Flesh-footed Shearwater 450 mm).
Abundance Common.
Status Native.

Family Procellariidae

Species *Puffinus griseus*

Common names
Sooty Shearwater, Muttonbird, Titi

Description
Upperparts Black.
Underparts Greyish-brown.
Underwing Primary feathers black with silver streaking on the secondary feathers.
Bill Dark grey.
Legs and feet Brownish.

Conspicuous features Smaller than the Flesh-footed Shearwater and Buller's Shearwater ✎ Silver streaking on the underwing ✎ Brownish-coloured feet separate it from the conspicuous flesh-coloured feet of the Flesh-footed Shearwater ✎ Dark-grey bill separates it from the pale-yellow bill of the Flesh-footed Shearwater.

Conspicuous characteristics A fast-flying shearwater with a rather stiff wing action ✎ Flaps and then glides for some distance along wave-tops and into the wave troughs.

Call 'Kuu-ah kuu-ah' heard from incoming birds over the breeding islands. Call can be hysterical in tone and deafening.

Breeding islands close to New Zealand Breeds on many islands from the Three Kings Islands to the north of the North Island, to the Campbell and Macquarie Islands to the far south of New Zealand. Very large populations breed on the islands around Stewart Island.

Breeding months November to May. One white egg is laid in a burrow, the first birds arriving back to the breeding colonies in late September.

Range worldwide Undertakes a circular migration of the Pacific Ocean, moving into the northern hemisphere waters around the first week of May.

Target localities Seas out from Sandspit in Northland ✎ Stirling Point, Bluff ✎ Seas around Southland and Stewart Island.

Flesh-footed Shearwater

Conspicuous colour Black.
Habitat Sea.
New Zealand range Ranges from Cook Strait northwards but more common in northern waters, especially in the Hauraki Gulf.
Size 440 mm (House Sparrow 145 mm, Black Petrel 460 mm).
Abundance Common.
Status Native.

Family Procellariidae **Species** *Puffinus carneipes*

Common names Flesh-footed Shearwater, Taonui

Description Upperparts Brownish black. Underparts Slightly paler than the upperparts. Bill Pale-yellow with top edge and tip dark grey. Feet Flesh-coloured.

Conspicuous features Dark-black plumage / Flesh-coloured feet.

Conspicuous characteristics Slow wing-flapping flight / Glides along waves with the occasional toss into the air / Often will alight and sit on water near fishing boats.

Call A wailing 'ku-koo-wah' heard after dark above nesting burrows.

Breeding islands close to New Zealand Breeds on many offshore islands from the Bay of Plenty northwards. Main colonies are on the Mercury Islands group and the Hen and Chicken Islands. It also breeds on islands in Cook Strait and off Taranaki on the Sugarloaf Islands. Also Lord Howe Island off Australia.

Breeding months November to April. One white egg is laid in a burrow.

Range worldwide New Zealand and Australian birds range north into the Pacific to Alaska and Siberia during their winter migration.

Target locality Seas out from Sandspit, Northland.

Flesh-footed Shearwater

Differences between the Flesh-footed Shearwater and the Sooty Shearwater

- Flesh-footed Shearwaters have darker plumage than Sooty Shearwaters and lack the silver underwing of the Sooty.
- Flesh-footed Shearwaters have yellowish bills; Sooty Shearwaters have dark-grey bills. (Bill colouring also separates it from Black Petrel, Westland Black Petrel and White-chinned Petrel.)
- Flesh-footed Shearwaters have pale-yellow bills with a grey top edge; Sooty Shearwaters have a dark-grey bill.
- Flesh-footed Shearwaters have flesh-coloured feet; Sooty Shearwaters have dark lilac-brown feet.

Sooty Shearwaters

Wedge-tailed Shearwater

Conspicuous colour Black.
Habitat Sea.
New Zealand range Found in the waters of northeastern New Zealand.
Size 460 mm (House Sparrow 145 mm, Flesh-footed Shearwater 450 mm).
Abundance A regular visitor to the northeastern waters off Northland.
Status Native.

Family Procellariidae

Species *Puffinus pacificus*

Common names Wedge-tailed Shearwater, Koaka

Description Upperparts and underparts Black. Underwing Black, with paler primaries. Bill Slate grey (shorter than Sooty Shearwater). Legs and feet Flesh-coloured.

Conspicuous features The long wedge-shaped tail is the most distinctive diagnostic feature ✎ Larger than Sooty Shearwater ✎ Lack of underwing silver separates it from Sooty Shearwater ✎ Legs are paler than the Flesh-footed Shearwater ✎ Slate-grey bill separates it from the pale-yellow bill of the Flesh-footed Shearwater and the darker grey of the Sooty Shearwater.

Conspicuous characteristics A fast flier ✎ Usually observed as a solitary bird, not in groups.

Call 'Ka woo ah' heard from incoming birds over the breeding islands.

Breeding islands close to New Zealand Breeds on Kermadec Islands to the north of New Zealand.

Breeding months December to June. One white egg is laid in a burrow the first birds arriving back at the breeding colonies in late October and early November.

Range worldwide Breeds from South Indian Ocean to New Zealand. New Zealand birds migrate in winter into the northeast Pacific.

Target locality Seas out from Sandspit in Northland.

Black Petrel

Conspicuous colour Black.
Habitat Sea.
New Zealand range Around the northern seas of the North Island and occasionally to Cook Strait.
Size 460 mm (House Sparrow 145 mm, Westland Black Petrel 480 mm, White-chinned Petrel 550 mm).
Abundance Rare but numbers increasing.
Status Endemic.

Family Procellariidae

Species
Procellaria parkinsoni

Common names
Black Petrel, Taiko

Description
Upperparts Black.
Underparts Black.
Bill Pale-yellowish with dark-grey tip.
Feet and legs Black.

Conspicuous features An all-black petrel ✎ Pale-yellowish bill with dark-grey tip.

Conspicuous characteristics Wheels, glides, tosses and skids down waves in rough weather ✎ Stiff-winged and continuous flier in calm weather.

Call Staccato-like with clacks and dull moans.

Breeding islands close to New Zealand Breeds on Little Barrier and Great Barrier Islands. Numbers have increased in both colonies since the control of cats and rats.

Breeding months November to July. One white egg is laid in a burrow.

Range worldwide Eastwards from New Zealand into the tropical Pacific and west across the Tasman to Australia.

Target locality Seas out from Sandspit, Northland.

Westland Black Petrel

Conspicuous colour Black.
Habitat Sea.
New Zealand range Western waters of the South Island to Taranaki and up the southeast coast of the North Island.
Size 480 mm (House Sparrow 145 mm, Black Petrel 460 mm, White-chinned Petrel 550 mm).
Abundance Uncommon.
Status Endemic.

Family Procellariidae Species *Procellaria westlandica*

Common names Westland Black Petrel, Westland Petrel

Description Upper and underparts Black. Bill Pale-yellowish with a black bill tip. Feet and legs Black.

Conspicuous features An overall black bird Pale-yellowish bill with dark bill tip.

Conspicuous characteristics Steady and even wing-beat in direct flight Over water flight involves wheeling, turning and skimming down waves.

Call Makes noisy guttural 'coo-coo-ra' noises near its burrow.

Breeding place The steep ranges in the Punakaiki district, near Greymouth in Westland, in the South Island.

Breeding months April to December. One white egg is laid in a burrow.

Range worldwide Migrates from its breeding colonies mainly east of New Zealand beyond the Chatham Islands and into the Tasman Sea towards Australia.

Target locality Punakaiki district, Westland.

White-chinned Petrel

Conspicuous colour Black.
Habitat Sea.
New Zealand range North to the Northland coast in winter and to Cook Strait in summer. Most commonly seen off the southeast coast of the South Island, from Otago south in summer.
Size 550 mm (House Sparrow 145 mm, Black Petrel 460 mm, Westland Black Petrel 480 mm).
Abundance Common locally south of Dunedin.
Status Circumpolar.

This bird shows no white

Family Procellariidae **Species** *Procellaria aequinoctialis*

Common name White-chinned Petrel

Description Upper and underparts Black. Chin Variable amounts of white. Bill Pale-yellow with dark grey at the end of nasal tubes in front of nostrils and on the groove of lower mandible.

Conspicuous features The pale-yellow bill lacks the dark tip of the Black Petrel and the Westland Black and also the dark top and tip of the bill of the Flesh-footed Shearwater The white chin is seldom seen on birds in New Zealand waters.

Conspicuous characteristics A slow and lazy flier Has a tendency to follow ships In rough weather tosses and wheels.

Call Near the burrow groans, squeals, clacks and clatters may be heard.

Breeding islands close to New Zealand Breeds on Auckland Islands, Campbell Island, Antipodes Islands and their surrounding islands. Also breeds on many islands around the southern oceans.

Breeding months November to May. One white egg is laid in a burrow.

Range worldwide Southern oceans from the Antarctic to about 30°S, although ranging further north in winter around southern Australia, South America and South Africa.

Target localities Otago Peninsula (Taiaroa Head and Cape Saunders), sometimes seen from the Royal Albatross observatory from mid-March onwards ✎ Nugget Point – often seen from near the Nugget Point lighthouse and from the Yellow-eyed Penguin lookout above Roaring Bay.

Locality and breeding differences between the White-chinned Petrel (*Procellaria aequinoctialis*), the Westland Black Petrel (*P. westlandica*) and the Black Petrel (*P. parkinsoni*)

- The Westland Black Petrel is a winter breeder, occupying its breeding colonies from February to December. The Black Petrel breeds from October through to July and the White-chinned Petrels breed from November to May.
- The Westland Black Petrel is considered to be non-migratory in the true annual migration sense, spreading across the Tasman towards Australia only. The Black Petrel migrates northeast of New Zealand and above the equator. The White-chinned Petrel is truly circumpolar.
- The breeding colonies of the three species are widely separated: the Black Petrel breeds on Little Barrier Island and Great Barrier Island; the Westland Black Petrel nests on the Paparoa Range of mainland South Island; the White-chinned Petrel nests well to the south of New Zealand on the Auckland Islands, Campbell Island and Antipodes Islands, as well as other islands around the southern oceans.

Similarities between the White-chinned Petrel, Westland Black Petrel and Black Petrel

- White-chinned Petrel: bill yellowish with dark grey only at the end of the nasal tubes. Pale-yellowish bill tip. Legs and feet, black.
- Westland Black Petrel: bill pale-yellowish with black tip. Legs and feet, black.
- Black Petrel: bill pale-yellowish with dark-grey tip. Legs and feet, black.
- Although the White-chinned Petrel is bigger than the other two, the three species are still difficult to separate at sea.
- The white chin of the White-chinned Petrel is not a good diagnostic feature as not all birds have it.

Southern Giant Petrel (dark phase)

Conspicuous colour Black.
Habitat Sea.
New Zealand range Around New Zealand for most of the year.
Size 900 mm (House Sparrow 145 mm, Cape Pigeon 400 mm).
Abundance Common.
Status Native.

Family Procellariidae Species *Macronectes giganteus*

Common names Giant Petrel, Nelly, Pungurunguru

Description – mature – dark phase Head Forehead pale-brown with sides of face and chin white or grey. Upperparts Black tending to brown. Underparts Tending to brown or grey. Underwings Brown with variable areas of grey. Bill Yellowish tipped green. Has a prominent nasal tube. Feet and legs Dark grey.

Description – mature – white phase Upperparts and underparts White with brown flecks.

Description – immature Upperparts and underparts Either white or black.

Conspicuous features Immature bird, either all black or all white ✎ Yellow bill tipped with green which separates it from the Northern Giant Petrel, which has a brown tip to end of bill ✎ Heavy nasal tube is noticeable.

Conspicuous characteristics Flight is often straight and direct ✎ It often glides and wheels behind ships ✎ Can glide motionless for some distance.

Breeding islands close to New Zealand Macquarie Island.

Breeding months October to April. Birds tend to nest in small, loose communities.

Range worldwide Around the Antarctic.

Target localities Cook Strait ✎ Kaikoura coastline ✎ Dunedin coastline south.

Northern Giant Petrel (immature)

Conspicuous colour Black.
Habitat Sea.
New Zealand range Around New Zealand for most of the year.
Size 900 mm (House Sparrow 145 mm, Cape Pigeon 400 mm).
Abundance Common.
Status Native.

Family Procellariidae **Species** *Macronectes halli*

Common names Giant Petrel, Nelly, Pungurunguru

Description – immature Upperparts and underparts Black.

Description – mature Head Forehead pale-brown with sides of face and chin grey. Upperparts Greyish-brown. Underparts Tending to dark grey. Underwings Brown with variable areas of grey. Bill Light tan with distinctive light-brown tip. Has a prominent nasal tube. Feet and legs Dark grey.

Conspicuous features Immature bird, black ✎ Mature bird brownish colouring with lighter grey around the face, throat and upper breast ✎ Light-tan bill with light-brownish tip. This separates it from the Southern Giant Petrel, which has a yellow bill tipped with green ✎ Heavy nasal tube is noticeable.

Conspicuous characteristics Flight is often straight and direct ✎ It often glides and wheels behind ships ✎ Can glide motionless for some distance.

Breeding islands close to New Zealand Chatham Islands, Stewart Island, Antipodes Islands, Campbell Island, Auckland Islands and Macquarie Island.

Breeding months August to February. Birds tend to nest in small, loose communities.

Range worldwide New Zealand west to South Africa.

Target localities Waters near Little Barrier Island ✎ Cook Strait: regularly seen from the Cook Strait ferry ✎ Kaikoura: regularly seen from both the Kaikoura coastline and from whale-watching excursions.

Saddleback

Conspicuous colours Black and chestnut.
Habitat A bird of both secondary and mature native forest.
Range Found on Hen Island and nine islands off the North Island, and on Mokoia Island in Lake Rotorua. Found on 11 islands off the South Island and Stewart Island and on Motuara Island in Queen Charlotte Sound, Marlborough.
Size 250 mm (House Sparrow 145 mm, Bellbird 200 mm).
Abundance Rare.
Status Endemic. The saddleback is a member of the family of New Zealand wattlebirds which include the Kokako and the extinct Huia. Originally widely spread in both islands, it became confined only to Hen Island off the Northland coast, and several small islands off the southwest coast of Stewart Island. Since 1964 it has been transferred to other islands after the clearance of introduced predators.

Family Callaeidae
Species *Philesturnus carunculatus*
Common names Saddleback, Tieke
Subspecies The North Island and South Island species of Saddleback represent two distinct subspecies. The North Island subspecies (*P. c. rufusater*) is characterised by a faint yellow band across the top of its chestnut-coloured saddle. The South Island bird (*P. c. carunculatus*) lacks this and is distinctive in that its immature birds lack the chestnut-coloured saddle in the juvenile phase. The absence of the saddle on the juvenile South Island bird initially had it classified as a separate species, known as the Jack Bird.

Description

Upperparts and underparts Glossy black all over. Wing-coverts and back Chestnut, with a thin buff line above on the North Island subspecies, which is absent on the South Island bird. Wing primaries Black. Rump and undertail Chestnut. Bill Black with orange-red wattles at the gape and below bill.

Conspicuous features

Chestnut saddle on mature birds of both subspecies, and on the North Island juvenile · Orange-red wattles below gape, especially in breeding plumage birds · Buff line above saddle on the North Island bird.

Conspicuous characteristics

Often seen on or near ground · Strongly territorial and will challenge intruders with song · Often seen running up branches · Generally very active and noisy among the vegetation.

Call Song, delivered by both birds, sometimes in duet, is a loud staccato 'whuu huhuhu hook' in one chattering phrase or variations of this. Notes are rhythmic and uttered in a seesawing fashion. Separate localities have distinctive dialects. Birds sing from daybreak till dusk throughout the year.

Nest In a tree, rock cavities or holes, usually close to the ground. The nest is made of twigs, grasses and leaves and lined with tree-fern hairs. Up to four white, blotched brown eggs are laid.

Target localities Tiritiri Matangi Island, near Whangaparaoa · Mokoia Island in Lake Rotorua · Kapiti Island north of Wellington · Motuara Island in Queen Charlotte Sound.

Tomtit (male)

Conspicuous colours Black and white.
Habitat Old forests, secondary forests and exotic pines; absent from scrublands.
Range Most large forested areas of New Zealand, including Stewart Island.
Size 130 mm (House Sparrow 145 mm, Grey Warbler 100 mm).
Abundance Regularly seen in selected localities.
Status Endemic.

North Island Tomtit

Family Eopsaltriidae

Species *Petroica macrocephela*

Common names Tomtit, Miromiro (North Island), Ngirungiru (South Island)

Subspecies Five: North Island (*P. m. toitoi*), South Island (*P. m. macrocephala*), Chatham Island (*P. m. chathamensis*), Snares Island (*P. m. dannefaerdi*), Auckland Island (*P. m. marrineri*).

Description — male bird Head, throat and upperparts Black. Frontal dot above bill White. Wings Black with a conspicuous white wingbar. Underparts Pure white on the North Island bird and white washed with yellow on the South Island, Stewart Island and Chatham Island birds. Tail Black with white edges.

South Island Tomtit

Description – female bird Upperparts Brownish. Frontal dot above bill Less conspicuous than on male bird and sometimes absent. Wingbar White, but less obvious than on male birds. Chin and breast Grey-brown. Underparts Greyish.

Conspicuous features Bigger and heavier than the Grey Warbler ✎ Has prominent eye ✎ White breast (North Island) or white breast with yellow wash (South Island).

Conspicuous characteristics Has a confiding and engaging nature ✎ Birds seem to arrive from nowhere ✎ Have the habit of peering down with head on one side ✎ Although usually to be found inside the forest, Tomtits will venture out for short spells from forest verges into cleared areas ✎ Often seen clinging on to the side of trunks with head down ✎ Birds often seek out people, look, and then fly off.

Call Song, delivered by the male bird, is a fragile 'swee sweedle sweedle sweedle'. Male birds in nesting territories can be very vocal.

Nest A bulky structure of twigs, fern, moss and cobwebs, lined with feathers and tree-fern hairs, on a branch fork or in a cavity on a bank. Up to five cream, yellow-speckled eggs are laid.

Target localities Tiritiri Matangi Island – recently released (2004) ✎ Rotorua exotic forests.

New Zealand Robin (male)

Conspicuous colours Black and white.
Habitat Native forests but seldom found in small remnants. Also found in the exotic pine forests of the central North Island.
Range In the North Island confined to central areas and Mokoia Island in Lake Rotorua. Also on Great Barrier Island, Little Barrier Island, Tiritiri Matangi Island and Kapiti Island. In the South Island absent from Canterbury Plains and Central Otago. Found on Stewart Island and Ulva Island.
Size 180 mm (House Sparrow 145 mm, Tomtit 130 mm).
Abundance Common in selected areas.
Status Endemic. The New Zealand Robin, along with the Tomtit, is closely related to the Australian robins of the genus *Petroica*. It is not closely related to the European Robin Red-breast (*Erithacus rubecula*). It has similar characteristics, though, of the large head, bold dark eye and the noiseless flight.

Family Eopsaltriidae

Species *Petroica australis*
Common names New Zealand Robin, Toutouwai

Subspecies North, South and Stewart Island subspecies, all black in colour: North Island Robin (*P. a. longipes*), South Island Robin (*P. a. australis*), Stewart Island Robin (*P. a. rakiura*).

Description – North Island Robin (male) Upperparts Black. Frontal dot above bill White. Upper breast Almost black. Lower breast and abdomen White.

Description – South Island Robin (male) Upperparts Black. Frontal dot above bill White. Breast and abdomen White with hint of yellow.

Description – female Upper and underparts Dark grey and drabber than male birds.

Conspicuous features Larger than the Tomtit / Bold eye / White breast, more conspicuous in the South Island bird / White frontal dot.

Conspicuous characteristics A confiding nature and will approach humans / A habit of just appearing / Scratches through forest floor leaf litter like a Blackbird.

Call The song, delivered by the male bird, is a ringing 'tueet tueet tueet tueet tooo' on a descending scale. Birds are more vocal in mid-morning. They also have a distinctive alarm call, often issued when standing on the ground, not unlike the alarm call of the introduced Blackbird. Birds sing throughout the year.

Nest A bulky cup of twigs, fern and moss bound with cobwebs and lined with feathers, wool or tree-fern hairs. Up to four cream, brown-speckled eggs.

Target localities Tiritiri Matangi Island near Whangaparaoa / Wenderholm Regional Park, north of Waiwera, North Auckland / Pine forests near Rotorua / Eglinton Valley near Te Anau / Ulva Island, Stewart Island.

Australian Magpie (mature)

Conspicuous colours Black and white.
Habitat Suburbs, parks and open countryside.
Range Throughout New Zealand except for Westland and Fiordland.
Size 410 mm (House Sparrow 145 mm, Rook 450 mm).
Abundance Common.
Status Introduced.

White-backed female

Family Cracticidae

Species *Gymnorhina tibicen*

Common name Magpie

Subspecies White-backed Magpie (*G. t. hypoleuca*) and Black-backed Magpie (*G. t. tibicen*). Most birds seen in New Zealand are White-backed Magpies.

Description – **White-backed Magpie** Head and underparts Black. Nape and mantle White. Back White. Wings Black, with conspicuous white patch on secondary feathers. Uppertail and undertail White with black tip. Bill White, with a black tip.

Description – **Black-backed Magpie** Head and underparts Black. Nape and mantle White.

White-backed male

Black-backed Magpie

Back Black. Wings Black, with conspicuous white patch on secondary feathers. Uppertail and undertail White with black tip. Bill White, with a black tip.

Description – female birds and juveniles Upperparts The white areas of the male birds are grey. Underparts Brownish-grey. Bill Brownish.

Conspicuous features Bold black and white colouring Heavy white bill.

Conspicuous characteristics Feeds on the ground in small parties of up to about 14 birds Aggressive if intruders (including humans) approach a nest.

Call A chuckle of bell notes.

Nest A platform of twigs and grasses, the cup lined with fine grass or wool. Up to four green, olive-blotched eggs are laid.

Target localities Gardens and parks.

White-winged Black Tern

Conspicuous colours Black and white.
Habitat Generally a coastal species which feeds over muddy estuary pools and takes mainly aquatic insects. Moves inland to lakes and wet areas near the coast.
Range Occasionally arrives at wet areas in both islands of New Zealand.
Abundance Uncommon.
Size 230 mm (House Sparrow 145 mm, White-fronted Tern 400 mm).
Status Migrant.

Family Laridae

Species *Chlidonias leucopterus*

Common name
White-winged Black Tern

Note Breeds in eastern Europe and Siberia and migrates south to Africa, southeastern Asia, Indonesia and Australasia.

Description – breeding plumage
Head, neck, back and underparts Black. Wings White, with black-tipped pale-grey primaries
Underwing Coverts black, with grey primaries black-tipped.
Tail White above and below.
Bill and legs Red.

Breeding plumage

Description – non-breeding
Head White forehead with indistinct partial black cap.
Upperparts White.
Wings White, with grey leading edges and primaries.
Rump and underparts White.
Bill and legs Black with reddish tinge.

Conspicuous features Smaller than Little and Fairy Terns ✎ Indistinct black cap, in non-breeding birds, separates it from Little Terns.

Conspicuous characteristics Feeds while hovering just above the water ✎ Very buoyant when in flight, almost dancing in the air ✎ Will alight and settle among other terns on sticks or posts.

Call A 'keet keet keet' sound.

Target localities Ahuriri Estuary, Hawke's Bay ✎ Bexley Wetland, Christchurch ✎ Wainono Lagoon, south of Timaru.

Pied Stilt

Conspicuous colours Black and white.
Habitat Harbours, estuaries, swamps and lake verges.
Range Throughout New Zealand but not Fiordland.
Size 350 mm (House Sparrow 145 mm, Pied Oystercatcher 460 mm).
Abundance Common in harbours and estuaries and inland wetlands.
Status Native.

Family Recurvirostridae **Species** *Himantopus himantopus*

Common names Pied Stilt, Australasian Pied Stilt, Poaka

Description – mature Back of neck and wings Black, separated by a white collar. Head and underparts White. Bill Black. Legs Pinkish.

Description – immature Head White, with varying amounts of smudgy black. Wings Black but paler than the adult bird. Underparts White.

Colour variations Some stilts have variable black colouration. These are either juvenile birds or hybrid birds resulting from interbreeding with Black Stilts.

Conspicuous features Generally an immaculate-looking black and white. Long pinkish legs.

Conspicuous characteristics Birds frequently fly at night, calling loudly. Birds harass and scold intruders who come near the nesting site.

Call A high-pitched 'yep, yep, yep' barking noise.

Nest Usually a colonial nester but often solitary, making a cup-shaped nest of grasses on a mound. Up to four greenish, blotched with brown eggs are laid.

Target localities Kawakawa Bay east of Papakura. Miranda coastline, Firth of Thames.

Black Stilt (immature)

Conspicuous colours Black and white.
Habitat Specific coastal and inland localities.
Range South Island (summer). Some migrate north to the North Island (winter).
Size 400 mm (House Sparrow 145 mm, Pied Stilt 380 mm).
Abundance Endangered.
Status Endemic.

Family Recurvirostridae **Species** *Himantopus novaezelandiae*

Common names Black Stilt, Kaki

Description – immature Upperparts White. Wings Black. Underparts White.

Conspicuous feature Pierce (1984) observed the progression of young Black Stilts from the nestling stage to the third summer. He recorded a succession of steps by which young birds progressed from being black-winged only, to black-flanked, to smudgy-breasted, to flecked with grey and finally to totally black.

Call A yapping 'yep yep yep' sound similar to the Pied Stilt.

Target localities Hauturu Road, Oparau, where it touches the Kawhia Harbour (winter only) Ohau River about 5 km south of Twizel (summer only).

Pied Oystercatcher

Conspicuous colours Black and white.
Habitat During the breeding months of August to January birds live in inland areas of Otago and Southland in the South Island, nesting along riverbeds, on pasture and among arable crops. A few, probably immature, birds always summer over on the traditional wintering grounds of the North Island. Away from the breeding months, from January onwards, they move north to coastal estuaries and harbours.
Range In non-breeding months from Bay of Plenty harbours northwards. In breeding months can be expected over much of the South Island but not Fiordland.
Size 460 mm (House Sparrow 145 mm, Red-billed Gull 370 mm).
Abundance Common.
Status Native.

Family Haematopodidae Species *Haematopus ostralegus finschi*

Common names Pied Oystercatcher, South Island Pied Oystercatcher, SIPO, Torea, Redbill

Description Upperparts Black. Underparts White. Bill, eyes and legs Red.

Conspicuous features Of slighter appearance than the Variable Oystercatcher Red bill, legs and eyes (See also next page.)

Call Noisy 'kleep kleep kleep'.

Nest A scrape on the ground. Two brown, blotched dark-brown eggs are laid.

Target localities Miranda, Firth of Thames in summer State Highway 6 from Invercargill to Te Anau in the breeding season.

Variable Oystercatcher (variable phase)

Conspicuous colours Black and white.
Habitat A bird of coastal areas only, which inhabits sandy beaches, mudflats and rocky promontories.
Range Found around Northland, the Bay of Plenty and down the east coast to Wellington and up to Wanganui, but with a gap around the Taranaki coast. Found around the South Island with gaps along the coasts of Canterbury and South Westland.
Size 480 mm (House Sparrow 145 mm, Pied Oystercatcher 460 mm).
Abundance Regularly seen in selected areas.
Status Endemic.

Family Haematopodidae **Species** *Haematopus unicolor*

Common names Variable Oystercatcher, Torea, Torea-pango, Black Oystercatcher (South Island)

Description – all black phase Upperparts and underparts Black. Bill, legs and eyes Red.

Description – pied phase Upperparts Black. Underparts White in varying quantities. Bill, legs and eyes Red.

Conspicuous features Of heavier appearance than the Pied Oystercatcher. Red bill, legs and eyes.

Call Similar to the Pied Oystercatcher – a melodic but aggressive 'keeleep keeleep'.

Nest A scrape on a sandy beach or rock shelf. Up to three brown, blotched dark-brown eggs are laid.

Target localities Northland beaches; Curio Bay, Southland.

Relationship to Australian Sooty Oystercatcher (*H. fuliginosus*)

The Variable Oystercatcher is a close relative of the Australian Sooty Oystercatcher (*H. fuliginosus*) but it differs in that it has three colour phases: black, variable amounts of white and distinctly pied not unlike the Pied Oystercatcher. Variable Oystercatchers of Southland and Stewart Island are all black.

Differences between the Pied Oystercatcher (*H. ostralegus*) and the Variable Oystercatcher (*H. unicolor*)

- Pied Oystercatchers have a clean-cut line between the white underparts and the black upperparts. This is lacking on the Variable.
- Pied Oystercatchers have two clean-cut white shoulder tabs which reach up into the black upperparts. This is lacking on the Variable.
- Pied Oystercatchers appear a neater and smaller bird.
- Pied Oystercatchers have a slimmer bill.

Characteristic differences between Pied and Variable Oystercatchers

- Pied Oystercatchers are inland nesters; Variable nest along the coast.
- Pied Oystercatchers start laying in August; Variable start nesting in mid-October with a second brood in December or early January.
- Pied Oystercatchers migrate north from the breeding grounds.
- Variable Oystercatchers form up into small coastal flocks not far from breeding areas. Some remain resident on their breeding territories from one year to the next.

Similar characteristics of Pied and Variable Oystercatchers

- Both birds have similar 'keeleep keeleep' type calls.
- Both species defend nesting territories with broken-wing type displays.
- Both species feed their young until the fledgling stage, being the only wading bird to actually feed its chicks.
- Both species have similar pale, buffish-brown eggs, marked with black dots and splotches. Clutch size in both species is usually three eggs for the first brood and two eggs for the second or replacement broods.

Little Shag (black-and-white phase)

Conspicuous colours Black and white.
Habitat Coastlines, estuaries, harbours, inland rivers and lakes.
Range Well-spread throughout New Zealand, including Stewart Island.
Size 560 mm (House Sparrow 145 mm, Pied Shag 810 mm).
Abundance Common.
Status Native.

Family Phalacrocoracidae

Species *Phalacrocorax melanoleucos*

Common names Little Shag, Kawau-paka (less common), Little Pied Shag or White-throated Shag

Phases Adult Comes in four distinct plumage phases: white-faced; white-faced plus a white-throat; white-faced and totally white underparts; white-faced and a smudgy underparts phase.
Juvenile Comes in two distinct plumage phases: black but separated from the Little Black Shag by a yellow bill (little Blacks have lead-coloured bills; a very small crest separates black-phase juvenile Little Shags from Little Black Shags); smudgy white face and all-white underparts (it also has a yellow bill).

Description Head White face and throat is common on all phases, the white colouring extending up the face to the crown with only a narrow belt of black being visible on the top of the head. Upperparts Black. Underparts See various phases above. Bill Yellow. Feet and legs Black.

Conspicuous characteristics Sits in social groupings on harbour and lake jetties and on lake shorelines, often with all phases including the fledglings in the one group ✎ Follows flight paths from breeding colonies to feeding grounds ✎ Tends to feed singly and not in packs as does the Little Black Shag ✎ Can spend up to 20 seconds under water in each dive for food ✎ Sits with wings open to dry them.

Call Guttural croaks and squeals at the nesting site only.

Nest Makes a twig nest in trees in loose colonies, sometimes in large numbers. Egg laying is well spread over the months of August until March. Up to four bluish-green eggs are laid.

Target localities Lake Rotorua beyond the Government Gardens, Rotorua ✎ Otago Harbour at Portobello.

Black-backed Gull

Conspicuous colours Black and white.
Habitat Around the coast and over pasture in spring.
Range Found throughout New Zealand and its offshore islands.
Size 600 mm (House Sparrow 145 mm, Red-billed Gull 370 mm).
Abundance Common throughout New Zealand.
Status Native.

Mature

Immature

Family Laridae

Species *Larus dominicanus*

Common names Black-backed Gull, Southern Black-backed Gull, Karoro, Dominican Gull (after the black-and-white colours of the Dominican friars). Colloquially known as Mollyhawk.

Description – mature Head, neck and underparts White. Wings and back Black with primaries edged with white. Bill, legs and feet Yellow.

Description – immature until year three years old Upperparts and underparts Dull brown streaked with dark brown. Bill and legs Brown.

Conspicuous features Juveniles are overall brown ✎ The biggest bird on beaches.

Call A clear yodelling sound of some volume.

Nest Nests in loose colonies on open pasture or around the coast. Nest is a mound of grass and sticks. Up to four greenish, brown-blotched eggs are laid.

Target locality Common on most beaches.

Stewart Island Shag (pied phase)

Conspicuous colours Black and white.
Habitat Coastal favouring rocky coastlines.
Range Around the coast of the South Island from Timaru to Stewart Island and into Fiordland.
Size 680 mm (House Sparrow 145 mm, King Shag 760 mm).
Abundance Common in localised places around the coast.
Status Endemic.

Family Phalacrocoracidae **Species** *Leucocarbo chalconotus*

Common names Stewart Island Shag, Bronze Shag, Kawau-mapua

Description – pied phase Head Black. Eye-ring Blue. Caruncles Orange-yellow, tending more to orange than the King Shag. Upperparts Black. Feathering has a noticeable sheen when in breeding plumage. Wings Black, with a slash of white on scapular feathers. Underparts White from the throat down. Bill Grey. Feet and legs Pink.

Description – bronze form Upperparts and underparts Black, with an iridescent green sheen to the feathering.

Description – intermediate phase Upperparts Black. Underparts White in varying amounts.

Conspicuous features All-black head / Blue eye-ring.

Conspicuous characteristics Often seen in groups on offshore rocks / Feeds in deeper water than Pied Shags / More confiding than the King Shag.

Call Silent, except when displaying at the colonies, when grunts are made.

Nest On the ground, on rock ledges, or among rock crevices where it builds a nest made of plant and vegetable material. Two pale-blue eggs are laid. It reaches its breeding peak in September–October.

Target localities Oamaru breakwater / Taiaroa Head near the Royal Albatross observatory / Whero Island off Stewart Island.

King Shag

Conspicuous colours Black and white.
Habitat Coastal, found around rock stacks off offshore islands.
Range Confined to rocky stacks and small islands at the entrances of Pelorus and Queen Charlotte Sounds.
Size 760 mm (House Sparrow 145 mm, Pied Shag 810 mm).
Abundance Rare.
Status Endemic.

Family Phalacrocoracidae

Species *Leucocarbo caruncalatus*

Common names King Shag, New Zealand King Shag, Kawau-pateketeteke, Rough-faced Shag

Note This shag is a close relative of the Stewart Island Shag (*L. chalconotus*). It is separated by its lack of a crest during the breeding season and by the fact that it comes only in a pied phase. The Stewart Island Shag has both a pied and bronze phase.

Description Head Black. Eye-ring Blue. Caruncles Yellow (rough patches of flesh on the sides of the face at the base of the bill). Upperparts Black. Wings Black, with a slash of white on the scapular feathers. Underparts White. Bill Grey. Feet and legs Pink.

Conspicuous features Lack of white on face separates it from the Pied Shag. Blue eye-ring.

Conspicuous characteristics When in flight head is held lower than the Pied Shag. Flies close to the water.

Call Usually silent unless displaying, when grunts and croaks are made.

Nest A colonial nester making a nest of seaweed and twigs on the ground. Up to three pale-blue eggs are laid. Breeding starts in April and continues to August.

Target localities White Rocks, to the northwest of the tip of Arapawa Island in Queen Charlotte Sound. Chetwode Islands, Sentinel Rock, Duffers Reef and Trio Islands at the entrance of Pelorus Sound.

Pied Shag

Conspicuous colours Black and white.
Habitat Coastal, preferring clean water.
Range Coastal from Raglan on the west coast north and around to Gisborne in North Island. In South Island found from Marlborough Sounds to Christchurch and around the Southland coast west to Fiordland. Also on Stewart Island.
Size 810 mm (House Sparrow 145 mm, Red-billed Gull 370 mm).
Abundance Common in localised places around the coast.
Status Native.

Family Phalacrocoracidae **Species** *Phalacrocorax varius*

Common names Pied Shag, Karuhiruhi

Description – mature Crown Black. Face and side of neck Pure white. Eye-ring Blue. Facial skin Yellow. Upperparts Black. Underparts White. Bill Grey. Feet and legs Black.

Description – juvenile Upperparts Brownish. Underparts Dirty white.

Conspicuous features Slightly smaller than the Black Shag but bigger than the Little Shag. Shorter tail than the Little Shag.

Conspicuous characteristics Coastal pohutukawa trees are a favourite roost. Regularly sits on rocks and more commonly on those surrounded by water. Will sit in the company of the Spotted Shags and Little Shags. Will sometimes sit in groups on sandy beaches in horizontal, duck-like posture with tails cocked vertically above their backs.

Call Grunts, croaks and squeals at the nesting site.

Nest A colonial nester, making a twig nest in a tree. Up to four greenish eggs.

Target localities Clarks Beach, Manukau Harbour, west of Papakura. Ulva Island, Stewart Island.

New Zealand Storm Petrel

Conspicuous colours Black and white.
Habitat Sea.
Range Waters around Little Barrier Island.
Size 180 mm (House Sparrow 145 mm, Fluttering Shearwater 300 mm).
Abundance Rare. Probably local waters only. A recently discovered storm petrel, as yet undescribed, but thought to be *Oceanites maorianus*, a species collected by the Astrolabe Expedition in 1829.
Status Endemic.

Family Oceanitidae

Species *Oceanites maorianus* (tentative)

Common name New Zealand Storm Petrel

Description Head, neck, throat and upper breast Black. Upperwings Sooty black with darker black primary feathers. Underwing Black with whitish inner. Rump and tail Black. Underparts White with black streaked abdomen markings. Legs and feet Black.

Conspicuous features Feet extend behind tail when in flight. Tail is square.

Conspicuous characteristics Dips constantly into the sea with feet touching water. Does not walk on the water as does the White-faced Storm Petrel.

Major breeding islands near New Zealand Unknown.

Breeding months Unknown.

Range worldwide Unknown.

Target localities In waters around Little Barrier Island.

Common Diving Petrel

Conspicuous colours Black and white.
Habitat Sea.
New Zealand range Around New Zealand.
Size 200 mm (House Sparrow 145 mm, Little Shearwater 300 mm).
Abundance Common.
Status Native.

Family Procellaridae **Species** *Pelecanoides urinatrix*

Common names Diving Petrel, Kuaka

Subspecies Four subspecies are recognised, of which only *P. u. urinatrix* is the most likely to be encountered on mainland New Zealand.

Description Upperparts Black. Face, neck and throat Mottled grey. Chin White. Underparts White. Bill Black. Legs and feet Blue.

Conspicuous features A small, short-tailed bird similar in size to a storm petrel ✎ Stump-tail is noticeable.

Conspicuous characteristics When sitting on the water it could be mistaken for a penguin ✎ Usually recognised by its fast flight ✎ Has a tendency to fly on a parallel plane straight into waves ✎ Stays under water for several seconds with each dive.

Call Very noisy near and at the breeding colony, making 'kuaka ka ka' type calls. The male and female calls can be recognised by experts.

Major breeding islands close to New Zealand Offshore islands from Three Kings Islands to Stewart Island, Chatham Islands and Snares Islands.

Breeding months August to February. One white egg is laid in a burrow.

Range worldwide Around New Zealand and towards Australia.

Target localities Tiritiri Matangi Island (near Whangaparaoa) waters ✎ Water towards Little Barrier Island from Sandspit ✎ Water around Stewart Island.

Little Shearwater

Conspicuous colours Black and white.
Habitat Sea.
New Zealand range Right around New Zealand.
Size 300 mm (House Sparrow 145 mm, Fluttering Shearwater 330 mm).
Abundance Common, especially in northern waters.
Status Endemic.

Family Procellariidae

Species *Puffinus assimilis*

Common names Little Shearwater, Allied Shearwater

Subspecies Of the seven recognised, the Kermadec Little Shearwater (*P. a. kermadecensis*), and the North Island Little Shearwater (*P. a. haurakiensis*) are most likely to be encountered.

Description Upperparts Black. Underparts White. Face White to just above the eye. Bill Dull blue with black ridge and tip. Feet Blue.

Conspicuous features The smallest shearwater in New Zealand waters · Bigger than a Diving Petrel and White-faced Storm Petrel · Dark in appearance · Blue feet.

Conspicuous characteristics Fast flying · Flight pattern is of several fast wing-beats followed by a short glide · Will skip and skim the waves in rough weather.

Call A rapid 'kakakakakaka...urr' is heard just on dusk near the breeding colony.

Major breeding islands close to New Zealand From Hen and Chickens Islands to the Mercury group off Coromandel Peninsula. The Kermadec Little Shearwater breeds on the Kermadec Islands.

Breeding months July to December. One white egg is laid in a burrow.

Range worldwide Stays close to the breeding colonies the year round.

Target localities Water towards Little Barrier Island from Sandspit · Waters east of Coromandel.

Cape Pigeon

Conspicuous colours Black and white.
Habitat Sea.
New Zealand range Around New Zealand.
Size 400 mm (House Sparrow 145 mm, Red-billed Gull 370 mm).
Abundance Common.
Status Native.

Family Procellariidae

Species *Daption capense*

Common names Cape Pigeon, Titore (also, less common, Cape Petrel, Pintado Petrel)

Subspecies Two are recognised, the Southern Cape Pigeon (*D. c. capense*) and the smaller Snares Cape Pigeon (*D. c. australe*). In winter and early spring both subspecies overlap in range and spread widely, well to the north of New Zealand.

Description Head and neck Black. Upperwings Black, lightly speckled with white, with two large, white patches on the upper primary feathers of each wing. Back and rump Mottled with black and white. Underwings White with black leading edges and thin, black trailing edges. Primary wing-tip feathers Black. Underparts White. Tail Mottled with black and white, and tipped with a broad, black tail-band. Bill and feet Black.

Conspicuous feature Four white wing patches.

Conspicuous characteristic Flies off the stern of boats and will follow ships for some distance.

Major breeding islands near New Zealand Breeds on Chatham Islands, The Pyramid (Chathams), Snares Islands, Antipodes Islands, Bounty Islands and Campbell Island.

Breeding months November to March. One white egg is laid in a scrape on a cliff.

Range worldwide Southern hemisphere to latitude 18°.

Target localities Waters out from Sandspit Waters in Cook Strait.

Yellow-nosed Mollymawk

Conspicuous colours Black and white.
Habitat Sea.
Range Seas from Bay of Plenty to Northland in winter.
Size 750 mm (House Sparrow 145 mm, Red-billed Gull 370 mm).
Abundance Uncommon but found occasionally in east coast waters in winter.
Status Circumpolar

Family Diomedeidae **Species** *Diomedea chlororhynchus*

Common names Yellow-nosed Mollymawk, Atlantic Yellow-nosed Mollymawk

Description Head White with greyish tinge to cheeks. Upperwings and back Black. Rump White. Underparts White. Underwing White with wing edges and wingtips black. Tail Black. Bill Black with upper edge of yellow, and orange-tipped.

Conspicuous features Underwing white finely edged with black. Smaller than both the Grey-headed and Buller's Mollymawks. Black bill with a yellow top edge. Buller's and Grey-headed have black-sided bills but with top ridges and bottom edges yellow.

Major breeding islands Islands in the South Indian Ocean and South Atlantic.

Breeding months September to April.

Range worldwide From the Tropic of Capricorn in the north to 55° south.

Target locality Waters off Tauranga.

Black-browed Mollymawk

Conspicuous colours Black and white.
Habitat Sea away from the coastline.
Range Around New Zealand.
Size 900 mm (House Sparrow 145 mm, Red-billed Gull 370 mm).
Abundance Common in northern waters.
Status Black-browed Mollymawk (*D. m. melanophrys*) circumpolar. New Zealand Black-browed Mollymawk (*D. m. impavida*) endemic.

Family Diomedeidae

Species *Diomedea melanophrys*

Common names Black-browed Mollymawk, Toroa

Subspecies Two are recognised: the Black-browed Mollymawk (*D. m. melanophrys*) and the New Zealand Black-browed Mollymawk (*D. m. impavida*).

Description — mature Head White, with a black triangle in front of and around eye. Upperwings and back Black. Neck and rump White. Underwings White, with a heavy black leading edge, wingtips and a lesser black trailing edge. Underparts White. Tail Dark grey with black tip. Bill Bright yellow with pink tip. Eye Honey-coloured with a surround of elongated black.

Description — immature Neck and nape Greyish. Underwing Greyish centres with black outer feathers.

Conspicuous features Black-browed Mollymawk (*D. m. melanophrys*) has about 50 percent of the white underwing edged heavily with black New Zealand Black-browed Mollymawk (*D. m. impavida*) has about 60 percent of the white underwing edged heavily with black.

Major breeding islands near New Zealand New Zealand Black-browed Mollymawk (*D. m. impavida*) breeds on Campbell Island. Black-browed Mollymawk (*D. m. melanophrys*) breeds on Macquarie Island, Antipodes Islands and Snares Islands.

Breeding months September to April.

Range worldwide From the Tropic of Capricorn in the north to 55° south.

Target localities Waters from Sandspit beyond Little Barrier Island Cook Strait waters Waters off Stewart Island.

Shy Mollymawk

Conspicuous colours Black and white.
Habitat Sea.
Range Around New Zealand.
Size 900 mm (House Sparrow 145 mm, smaller than a Royal Albatross but larger than Buller's Mollymawk).
Abundance Common especially in southern waters.
Status Native.

Family Diomedeidae

Species *Diomedea cauta*

Common names Shy Mollymawk (a name acquired from its habit of being shy about gliding behind ships), White-capped Mollymawk

Subspecies These include the Shy Mollymawk (*D. c. cauta*), Salvin's Mollymawk (*D. c. salvini*) and the Chatham Island Mollymawk (*D. c. erimita*).

Description – mature Head White with a black shadow in front of eye. Sometimes grey mottling on cheeks. Upperwing Black. Back Grey-black. Underwing White, with thin, black edging and black primary tips. Neck White. Rump White. Underparts White. Tail White and tipped with a broad band of dark grey. Bill Grey with yellowish top and tip.

Conspicuous features Whitish head cap ✎ Some birds have greyish cheeks ✎ Underwing is clean white with narrow black edges ✎ Length of wing appears longer than other mollymawks.

Conspicuous characteristics Has a habit of coming in behind fishing boats and looking for fish scraps ✎ Will alight on the water and feed.

Major breeding islands near New Zealand Auckland Islands and Antipodes Islands.

Breeding months November to August.

Range worldwide Around New Zealand to South Africa and the Indian Ocean.

Target localities Cook Strait ✎ Waters off Stewart Island.

Wandering Albatross

Conspicuous colours Black and white.
Habitat Sea.
Range Around New Zealand. More common in New Zealand waters in winter months.
Size 1150 mm (House Sparrow 145 mm, Shy Mollymawk 900 mm).
Abundance Uncommon.
Status Circumpolar.

Family Diomedeidae

Species *Diomedea exulans*

Common names Wandering Albatross

Subspecies Two are recognised as breeding in New Zealand: Antipodes Wandering Albatross (*D. e. antipodensis*), which breeds on Antipodes and Campbell Islands, and Gibson's Wandering Albatross (*D. e. gibsoni*), which breeds on the Auckland Islands. The Snowy Albatross (*D. e. chionoptera*), which breeds on Macquarie Island, is also a regular visitor to New Zealand waters.

Description – mature Head White. Upperwing White darkening to black on secondaries and primaries.
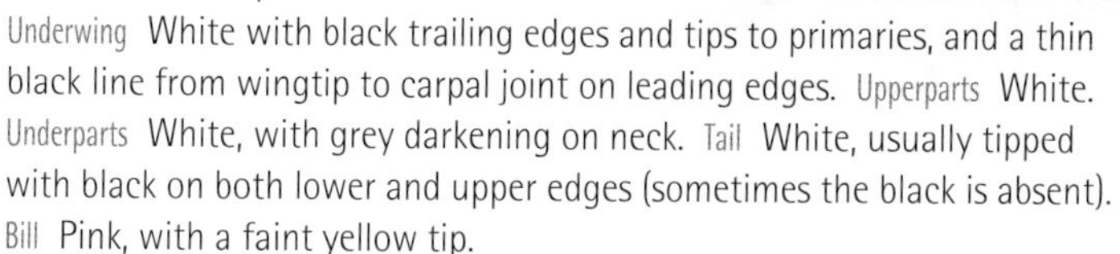
Underwing White with black trailing edges and tips to primaries, and a thin black line from wingtip to carpal joint on leading edges. Upperparts White. Underparts White, with grey darkening on neck. Tail White, usually tipped with black on both lower and upper edges (sometimes the black is absent). Bill Pink, with a faint yellow tip.

Description – immature Face White. Upperparts Pure black but with the black slowly being replaced by white with maturity. Underwing White with primaries on trailing edges lined with black. Underparts Black. Tail White, tipped with various amounts of black.

Breeding islands close to New Zealand Antipodes Island, Auckland Islands, Macquarie Island and Campbell Island.

Breeding months Starts in January and young birds leave the nest one year later. The parental pair then has one year's rest before commencing egg laying again.

Range worldwide The southern oceans north to the Tropic of Capricorn, occasionally straggling further north.

Target localities Waters from Sandspit beyond Little Barrier Island in North Island. Waters from Kaikoura in northern South Island.

Royal Albatross

Conspicuous colours Black and white.
Habitat Sea.
Range Around New Zealand but more regularly from Stewart Island to Cook Strait.
Size 1150 mm (House Sparrow 145 mm, Shy Mollymawk 900 mm).
Abundance Common.
Status Endemic.

Family Diomedeid

Species *Diomedea epomophora*

Common names
Royal Albatross, Toroa-whakaingo (Also called Northern Royal Albatross and Southern Royal Albatross)

Subspecies
Two are recognised: the Northern Royal (*D. e. sandfordi*) and the Southern Royal (*D. e. epomophora*).

Description – mature (Northern Royal)
Upper and underparts White. Upperwings Black. Underwings White, with black leading edges from carpal joint to wingtip and black along the trailing edge. Tail White, but sometimes black dotted on the undertail. Bill Pinkish, with cream tip and black cutting-edge to upper mandible.

Description – mature (Southern Royal) Upper and underparts White. Upperwings Black but with broad areas of white on inner leading edges. Underwings White with black leading edges from carpal joint to wingtip and black along the trailing edge. Tail White. Bill Pinkish with cream tip and black cutting-edge to upper mandible.

Description – immature Upperparts Traces of black on head, back and uppertail. Otherwise white. Northern Royals have more black markings than Southern Royals.

Breeding islands close to New Zealand Islands off the Chatham Islands, Auckland Islands, Enderby Island, Campbell Island, and on the South Island of New Zealand at Taiaroa Head on the Otago Peninsula.

Breeding months October to September. Breeding season commences about two months earlier than the Wandering Albatross.

Range worldwide Probably similar to the Wandering Albatross but does not wander as far north into the tropics.

Target localities Waters from Sandspit beyond Little Barrier Island in North Island
- Waters from Kaikoura in northern South Island.

Differences between Wandering and Royal Albatrosses

- Wandering Albatrosses usually show some breast darkening; Royals (Northern and Southern) have totally white breasts and abdomen.
- Wandering have black tips to undertail feathers; Royal (Northern) have some black undertail markings; Royal (Southern) have no black undertail markings.
- Wandering have black tips to uppertail; Royals have no black uppertail tips.
- Wandering have some white on primary and secondary upperwing feathers. This is similar to the Southern Royal. However Northern Royals have totally black upperwings.
- Northern Royals have very dark leading edges to the underwing, with the black extending heavily from end of wing to carpal joint.
- Juvenile Wandering Albatrosses usually have large amounts of upper black markings, traces of black around the head and upper back, and blotchy black and white upperwing markings.

Differences between Northern and Southern Royal Albatrosses

- Northern Royal adults have all black upperwings.
- Southern Royal adults have black upperwing with broad areas of inner white on the leading edge and across the back.

Australian Magpie (female, immature)

Conspicuous colours Black, white and grey.
Habitat Suburbs, parks and open countryside.
Range Throughout New Zealand except for Westland and Fiordland.
Size 410 mm (House Sparrow 145 mm, Rook 450 mm).
Abundance Common.
Status Introduced.

Family
Cracticidae

Species
Gymnorhina tibicen

Common names
Magpie, Australian Magpie

Description
Head and underparts Black. Nape and mantle White. Upperparts Grey. Underparts Brownish-grey. Wings Black, with a conspicuous white patch on secondary feathers. Uppertail and undertail White with black tip. Bill White on female, brownish on juvenile, with a black tip.

Conspicuous features Bold black-and-white colouring.

Conspicuous characteristics Feeds on the ground in small parties of up to about 14 birds Aggressive if intruders (including humans) approach a nest.

Call A chuckle of bell notes.

Nest A platform of twigs and grasses, the cup lined with fine grass or wool. Up to four green, olive-blotched eggs are laid.

Target localities Gardens and parks.

Stitchbird (male)

Conspicuous colours Black, white and yellow.
Habitat A bird of both secondary and mature native forest.
Range Little Barrier Island, Tiritiri Matangi Island and Kapiti Island.
Size 180 mm (House Sparrow 145 mm, Bellbird 200 mm).
Abundance Rare.
Status Endemic.

Family Meliphagidae

Species *Notiomystis cincta*

Common names Stitchbird, Hihi

Description – male Head Velvet black with white erectile tufts behind eyes. Upper breast and back Black. Wings Black with light-brown edges and a conspicuous white wingbar at base of primaries. Breast A band of yellow crosses the breast and across the wings below the black. Rump and underparts Pale brown.

Description – female Upper and underparts Olive-brown, similar to the female Bellbird. Wings Brown with a white wingbar.

Conspicuous features Brightly coloured White wingbar separates the female Stitchbird from the female Bellbird Has whiskers at the gape.

Conspicuous characteristics Birds are usually in pairs In the nesting season they are strongly territorial Out of the nesting season often follow flocks of Whiteheads through the forest Have a habit of alighting on a branch with the tail held high or even over backs or with heads facing down and tail kinked.

Call Has 'pek pek pek' alarm notes similar to the Bellbird and a loud 'zee zi ip'.

Nest Usually in holes made of sticks. Four white eggs are laid.

Target locality Tiritiri Matangi Island.

Pacific Golden Plover (breeding)

Conspicuous colours Black, white and yellow.
Habitat Coast and estuaries.
Range Scattered about New Zealand.
Size 250 mm (House Sparrow 145 mm, Wrybill 200 mm).
Abundance Common in selected areas.
Status Migrant.

Family Charadriidae
Species *Pluvialis fulva*
Common names
Golden Plover,
Pacific Golden Plover
Description
– breeding plumage
Crown of head and upperparts Brown, heavily flecked with golden-yellow and white. Neck A white dividing line, starting at the forehead and travelling behind the eye and down the neck, separates the black underparts from the golden upperparts. Underparts Black.

Description – non-breeding Head Brown with a white stripe above the eye. Throat Buff. Upperparts Brown with feathers edged with golden-buff. Breast Light brownish. Abdomen Brown but white undertail. Bill Black with white feathers around the base.

Conspicuous features The golden tonings ✎ Striking black underparts with white line from head to flanks.

Conspicuous characteristics Holds head high ✎ Runs and pauses when feeding ✎ Will feed on pasture ✎ Has an alert stance ✎ At high tide birds will stand motionless for some time ✎ Upon landing, birds hold their wings erect for a moment before folding ✎ Easily confused with the American Golden Plover.

Call Usually a clear two-syllabic 'tuill tuill' sound.

Target localities Aroha Island, Kerikeri ✎ Jordans Road, Kaipara Harbour, north of Helensville ✎ Clarks Beach and Te Hihi, Manukau Harbour ✎ Waituna Lagoon, east of Invercargill.

Differences between the Pacific Golden Plover (*P. fulva*) and the American Golden Plover (*P. dominica*)

See Pacific Golden Plover – non-breeding (p. 177).

Fiordland Crested Penguin

Conspicuous colours Black, white and yellow.
Habitat Coast and fiords.
Range Around the South Island.
Size 600 mm (House Sparrow 145 mm, Blue Penguin 400 mm).
Abundance Uncommon.
Status Endemic.

Family Spheniscidae **Species** *Eudyptes pachyrhynchus*

Common names Fiordland Crested Penguin, Tawaki, Pokotiwha

Other species The Fiordland Crested Penguin is closely related to the Erect-crested Penguin (*E. sclateri*), which breeds well to the south on the Bounty and Antipodes Islands, and the Snares Crested Penguin (*E. robustus*), which breeds on the Snares Islands.

Description Head Glossy black to under throat. Eyebrow A yellow band goes from above the eye to the back of head. Crest Yellow, slightly drooping down the neck. Upperparts Soft black. Underparts White from throat down. Flippers Soft black, with a faint white trailing edge. Bill Reddish-brown.

Conspicuous features Yellow crest, but no yellow-tinged head like the Yellow-eyed Penguin ✎ Underparts are all white ✎ Lacks the flesh-coloured gape of the Snares Crested Penguin.

Conspicuous characteristics Will sit on rocks in the sun on the vegetation line ✎ Gives itself away with vocal displays of grunting, mewing and growling near the nesting sites ✎ Heard more during morning hours.

Call A loud braying sound.

Nest Returns to nesting areas in July. Nests are in cavities and under vegetation near the shoreline. Birds are isolated nesters but nests are usually within calling distance of other pairs. Two white eggs are laid in early August and birds fledge at the end of November. Fiordland Crested Penguins nest about two months earlier than Snares Crested Penguins.

Moult Returns to nesting grounds to moult during January and February.

Target localities Knight Point near Lake Moeraki, Westland ✎ Milford Sound, Fiordland ✎ Paterson Inlet, Stewart Island.

Differences between the Fiordland Crested Penguin, Erect-crested Penguin and Snares Crested Penguin

- Fiordland Crested lacks the more erect tufted-type crest of the other two species. Its crest is broad but not tufted.
- Fiordland Crested lacks the fleshy gape marking which is noticeable on Snares Crested.
- Fiordland Crested tends to have a finer bill and head.
- Fiordland Crested has slightly lighter cheek colouration when compared with the black of the other two species.

Yellow-eyed Penguin

Conspicuous colours Black, white and yellow.
Habitat Coast and harbours.
Range East coast of the South Island from Banks Peninsula south. Also around Stewart Island and on Campbell and Auckland Islands.
Size 650 mm (House Sparrow 145 mm, Blue Penguin 400 mm).
Abundance Uncommon.
Status Endemic.

Family Spheniscidae

Species *Megadyptes antipodes*

Common names Yellow-eyed Penguin, Hoiho

Description Head Yellowish, streaked with black. A broad yellow band passes through the eye and around the crown. Upperparts Soft black. Flippers Upper surface black, edged with white. Underparts White. Eye Yellow.

Conspicuous features Larger than the Fiordland Crested Penguin ✎ Streaked yellow head ✎ Yellow eye.

Conspicuous characteristics Comes ashore to breeding colonies from early afternoon ✎ Stands for some time after landing on the beach ✎ Only comes ashore on sand rather than rocks ✎ Surfaces when swimming regularly but only for short intervals.

Call A musical trumpet heard from birds at the nesting sites.

Nest Starts visiting nesting sites in July with egg laying commencing in September. Fledglings leave the nest during the following April. Breeds in loose colonies on vegetated dunes with nests being built under vegetation and well spaced but within calling distance of each other. Two bluish-green eggs are laid.

Target localities Bushy Beach south of the Oamaru Harbour breakwater ✎ Sandfly Bay, Otago Peninsula ✎ Roaring Bay, Nugget Point, South Otago.

Kingfisher

Conspicuous colour Blue.
Habitat Parks, gardens, open country, coast and forests.
Range Throughout New Zealand but less common in Southland.
Size 240 mm (House Sparrow 145 mm).
Abundance Common.
Status Native.

Family Alcedinidae **Species** *Halcyon sancta*

Common names Kingfisher, Kotare, New Zealand Kingfisher, Sacred Kingfisher

Description Cap Blue-green. Face A buff line extends above eye from bill to just beyond eye. Around the eye is black. Behind eye is blue. Neck A white collar extends from chin to nape. Upperparts Greenish-brown, with deep-blue wing primaries. Underparts Orange-tan. Tail Deep blue. Bill Black.

Conspicuous features Blue head, wings and tail • Heavy, black bill.

Conspicuous characteristics Sits on power wires, fences, or high up on the branches of dead trees • Sits on foreshore rocks • Commonly sits in trees above rivers or lakes • Takes insects or worms from the ground or from water.

Call A continuous and often monotonous 'ki-ki-ki-ki'. At times, when in full song, a syllable is added. Near the nest it has a guttural screech.

Nest A hole in a clay bank or old tree. Up to five white eggs are laid.

Target localities Common in most areas but not in Southland • Miranda, Firth of Thames.

Pukeko

Conspicuous colour Blue.
Habitat Open country, pasture, swamps.
Range Throughout New Zealand.
Size 510 mm (House Sparrow 145 mm).
Abundance Common.
Status Native. A cosmopolitan species well-spread through parts of southern Europe, Asia, Africa, Australia, New Zealand and many of the Pacific islands.

Family
Rallidae

Species *Porphyrio porphyrio*

Common names
Pukeko, Swamp Hen

Description
Upperparts Black.
Underparts Purple.
Undertail White.
Bill and frontal shield Red.
Legs and Feet Orange-red.

Conspicuous features
Purple breast and underparts ✎ White undertail.

Conspicuous characteristics Flies with rapid wing-beats and reddish legs trailing ✎ If alarmed, runs rapidly through grass or rushes with head down and with white tail patch showing ✎ Birds when in territories during the breeding season noisily challenge and fight birds from neighbouring territories.

Call A variety of harsh screech sounds and a booming 'pu-ku pu-ku'. Also a rapid 'kuweek kuweek'.

Nest A platform of grasses among rushes. Up to ten buff, brown-blotched eggs are laid. Birds usually form pairs but polygamous relationships are sometimes established, with all birds sharing incubation and rearing.

Target localities Common throughout the countryside ✎ Shakespear Regional Park, Whangaparaoa Peninsula.

Takahe

Conspicuous colour Blue.
Habitat High-country, tussock areas.
Range Murchison Range near Te Anau. Some offshore islands.
Size 630 mm (House Sparrow 145 mm, Pukeko 510 mm).
Abundance Rare. Was thought extinct but rediscovered in 1948 by Dr G.B. Orbell in the Murchison Mountains above the shores of Lake Te Anau on its western side.
Status Endemic.

Family Rallidae **Species** *Porphyrio mantelli*

Common names Takahe, South Island Takahe (formerly Notornis, from *Notornis mantelli*)

Description Head Indigo-blue. Upperparts Olive-green. Wings Olive-green. Underparts Indigo-blue. Undertail White. Bill and frontal shield Red. Legs Red.

Conspicuous features Large size, but similar to a Pukeko ✎ Heavy, red bill.

Conspicuous characteristics Slow-moving when feeding ✎ Cumbersome gait.

Call Slow deep 'coo eet coo eet' sounds.

Nest On the ground among tussock. Two pale-buff eggs are laid.

Target localities Tiritiri Matangi Island ✎ Kapiti Island.

Peafowl (Peacock)

Conspicuous colour Blue.
Habitat Parks, gardens and open country.
Range Selected areas in Northland, South Auckland, Bay of Plenty, Poverty Bay, Hawke's Bay, Taranaki, northwest South Island and North Canterbury.
Size 900 mm (House Sparrow 145 mm, Arctic Skua 430 mm).
Abundance Uncommon.
Status Introduced.

Family Phasianidae

Species *Pavo cristatus*

Common name Peafowl

Description – male

Crown, neck, breast and inner-wing primaries Iridescent blue. A tuft of black fan-like feathers, tipped with blue, protrude from crown. Face White with a black stripe through eye. Chin Green. Back Gree n. Wings Tightly barred with buff, black and chestnut. Underparts Iridescent green. Tail The most conspicuous part, especially in the mating season. When raised it forms a spreading fan of long, green-coloured feathers studded with circular whirls of blue and brown.

Description – female

Crown and neck Metallic green. A tuft of fan-like feathers, tipped with green, protrude from the back of head. Face, chin and throat White with brown eye-stripe. Upperparts Brown. Breast Buff-coloured. Underparts Whitish.

Conspicuous feature Head crest.

Conspicuous characteristics

Noisy at dusk when they gather together in trees for the night ✎ A gregarious nature ✎ Lost birds become agitated, calling constantly.

Call Day call is a loud 'may-wee' screech. Evening call is more of a wail.

Nest A bowl in grasses or low vegetation. Up to six cream eggs are laid.

Target localities Waitangi Forest near Paihia in Northland ✎ Shakespear Regional Park at the end of Whangaparaoa Peninsula, North Auckland.

Welcome Swallow

Conspicuous colours Blue, black, red.
Habitat Pasture, open country and wetlands.
Range Throughout New Zealand.
Abundance Common arriving in New Zealand from Australia in the 1950s when it established a small breeding colony in the far north from where it populated the whole country.
Size 150 mm (House Sparrow 145 mm).
Status Native.

Family Hirundinidae **Species** *Hirundo neoxena*

Common name Welcome Swallow

Description Crown Blue-black. Forehead, throat and breast Rusty-brown Upperparts Metallic blue-black. Underparts Greyish-white.

Conspicuous features Rusty-brown of the chin and throat ✎ Forked tail.

Conspicuous characteristics When feeding over water birds continually dip into the water surface ✎ In winter it can be seen feeding over dams and ponds in large flocks ✎ Often seen sitting along power wires in large numbers, usually towards the end of the breeding season.

Call A short 'peep peep' is heard when birds are feeding. Sometimes a rapid twittering is uttered from a perch.

Nests These are found under bridges, eaves, commonly in milking sheds, as well as along cliffs and banks of rivers. They are cup-shaped and carefully sculptured with mud pellets made of clay mixed with saliva. They are lined with straw, feathers and wool. Up to four pink, brown-flecked eggs are laid.

Target localities Common everywhere away from alpine areas.

California Quail (male)

Conspicuous colours Blue and grey.
Habitat Pasture, and wooded areas.
Range Throughout the North Island, and the South Island except for Southland, Fiordland and forested areas of the West Coast.
Size 250 mm (House Sparrow 145mm, Mallard 580 mm).
Abundance Common.
Status Introduced.

Family Phasianidae

Species *Callipepla californica*

Common name California Quail

Description – male Forehead Grey to the eye. Crown Black, with a white edge band and a black crest feather which leans forward. Face and chin Black edged with a white-curved throat collar. Nape Black-and-white spotted. Wings Dark grey. Breast Blue-grey. Underparts Buff, feathers edged with black to give a horizontal scaly effect. Chestnut hue to middle abdomen. Bill Black. Legs and feet Grey.

Description – female Head, breast and upperparts Duller than the male and tending to brown. Crest feather Shorter than male. Underparts Similar to male.

Conspicuous features Black and white on the head of the male bird ✎ Topknot feather ✎ Scaly, buff-coloured abdomen.

Conspicuous characteristics Usually to be found in pairs or coveys ✎ When disturbed birds run and then fly off with whirr of wings ✎ Male birds will call for lengthy periods from a conspicuous perch.

Call A noisy tri-syllabic 'ki kuu kuu'.

Nest A hollow in grasses. Up to 13 cream, brown-blotched eggs are laid. A late breeder with chicks being hatched from mid-December onwards.

Target localities Aroha Island near Kerikeri ✎ Common in areas of scrubland and wilderness.

Chukor

Conspicuous colours Blue and grey.
Habitat Dry high-country areas of South Island.
Range Marlborough to Central Otago.
Size 310 mm (House Sparrow 145 mm, California Quail 250 mm).
Abundance In selected areas.
Status Introduced.

Family Phasianidae **Species** *Alectoris chukar* **Common name** Chukor

Note A native of India introduced into New Zealand in 1926 as a game bird.

Description Head and nape Grey tinged with red, with a faint white stripe above eye. Face and chin White, the white ringed with a narrow black band. Upperparts Bluish-grey. Underparts Grey tinged with brown. Sides and flanks Vertically barred with alternate black, white and chestnut lines. Bill Red. Legs and feet Pink.

Conspicuous features Black, white and chestnut vertical barring on sides and flanks ✎ White face with frame of black.

Call A carrying, high-pitched 'chuck chuck chuck chukar'.

Nest A hollow in grasses or tussock on the ground. Up to 14 cream, brown-blotched eggs are laid.

Target localities Crown Range, Central Otago ✎ The Mount John Observatory, Lake Tekapo ✎ Lake Pukaki, South Canterbury – the road to Mount Cook.

Rock Pigeon

Conspicuous colours Blue and grey.
Habitat Cities, parks, pasture and mudflats.
Range Throughout both the main islands, usually close to cities.
Size 330 mm (House Sparrow 145 mm, California Quail 250 mm).
Abundance Common mainly in cities.
Status Introduced.

Family Columbidae **Species** *Columba livia*

Common name Homing Pigeon

Description Head and upperparts Variations of dark grey and blue-grey with some purple iridescence on neck. Wings Usually blue-grey with two black bars on primaries. Rump White.

Conspicuous features Two prominent blacks bars on wings ✎ Comes in a variation of colours.

Conspicuous characteristics Confiding in cities ✎ Feeds on the ground ✎ Breeds high up on building ledges or under bridges ✎ Has established populations on cliff faces in places like Orewa, Napier, Rangitikei and inland Canterbury.

Call 'Brruuu brruu brruuu brruu' or 'ooo rooo coo' sounds.

Nest A rough platform of twigs and grasses. Two white eggs are laid.

Target localities Orewa Beach, north end ✎ Cliffs above Napier ✎ Many cities and towns.

Nankeen Kestrel (male)

Conspicuous colours Blue and grey.
Habitat Pasture and wooded areas.
Range Selected areas.
Abundance Rare but a regular visitor.
Size 330 mm (House Sparrow 145mm, New Zealand Falcon 430 mm).
Status Native.

Family Falconidae **Species** *Falco cenchroides*

Common names Nankeen Kestrel, Australian Kestrel

Description – male Head and neck Blue-grey. Ceres (above bill) Yellow. Back and upperwings Cinnamon-brown with black primaries. Underparts White, with wash of buff on upper breast. Tail Grey with a black tip. Eye Brown, with a thin, yellow eye-ring. Bill Tan-coloured with a black tip. Legs Yellow with black claws.

Description – female Head Rufous with some black streaking. Upperparts Cinnamon-brown, feathers spotted or streaked black. Underparts Light brown breast to white abdomen. Tail Brown, barred with ten rows of black and a black tip to tail.

Conspicuous features Small size Blue grey head and tail.

Conspicuous characteristics Hovers in flight above prey before gliding down Tends to fly with fast wing-beats and then soars with flat wings.

Call Shrill, excited chatter.

Nest Rock cavities, on cliff ledges or in old trees. Up to five buff, reddish-blotched, eggs are laid. Nesting has not been recorded in New Zealand.

Target localities None, but has been seen near Naike, North Waikato and Te Mata Peak, Havelock North, Hawke's Bay.

Kokako

Conspicuous colours Blue and grey.
Habitat Back-country forests.
Range Puketi Forest (Kerikeri district), central North Island forests and offshore islands.
Size 380 mm (House Sparrow 145 mm).
Abundance Rare.
Status Endemic.

Family Callaeidae

Species *Callaeas cinerea*

Common names
Kokako; North Island Kokako or Blue-wattled Crow (North Island), Orange-wattled Crow (South Island, possibly extinct).

Description
Forehead and face Covered with a black mask to behind the eye.
Wattles Found just below gape, are blue on the North Island bird and orange on the South Island bird.
Upperparts Grey.
Wings Grey with a touch of black on edges of primaries.
Underparts Grey.
Bill and feet Glossy black.

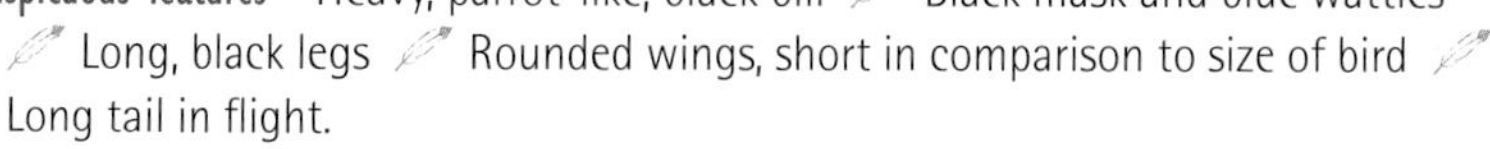

Conspicuous features Heavy, parrot-like, black bill ✎ Black mask and blue wattles ✎ Long, black legs ✎ Rounded wings, short in comparison to size of bird ✎ Long tail in flight.

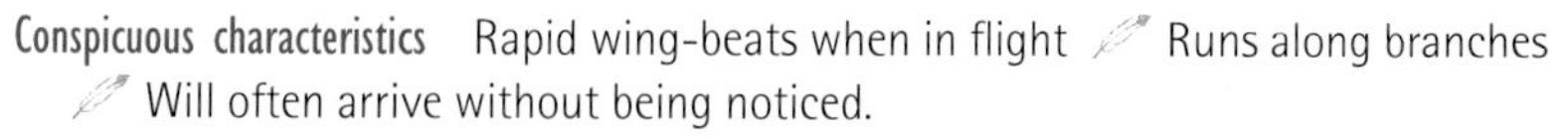

Conspicuous characteristics Rapid wing-beats when in flight ✎ Runs along branches ✎ Will often arrive without being noticed.

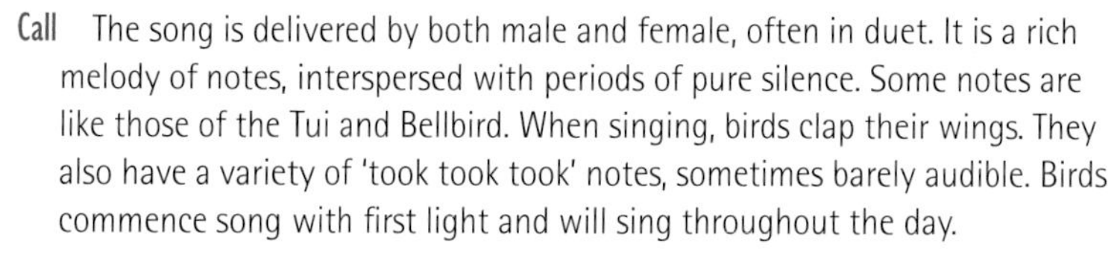

Call The song is delivered by both male and female, often in duet. It is a rich melody of notes, interspersed with periods of pure silence. Some notes are like those of the Tui and Bellbird. When singing, birds clap their wings. They also have a variety of 'took took took' notes, sometimes barely audible. Birds commence song with first light and will sing throughout the day.

Nest A platform of twigs and leaves, the cup lined with tree-fern hairs or fine grasses at about 4 m from the ground. Up to four grey, brown-blotched eggs.

Target localities Tiritiri Matangi Island near Whangaparaoa ✎ Pureora Forest near Benneydale, King Country.

New Zealand Shoveler (male)

Conspicuous colours Blue, grey and brown.
Habitat Ponds and lakes, especially those with indented rush and raupo-covered shorelines; also slow-moving rivers, estuaries and muddy coastal strips.
Range Evenly spread through the North and South Islands. Reaches some of the higher altitude lakes, especially where seclusion is guaranteed.
Size 490 mm (House Sparrow 145 mm, Mallard 580 mm).
Abundance Common.
Status Endemic.

Family Anatidae

Species *Anas rhynchotis*

Common names Shoveler, Spoonbill, Kuruwhengi

Description – male (breeding plumage) Head and neck Bluish-grey, with black behind bill and a white crescent-shaped stripe in front of a golden eye. Back Brown. Wings Bluish-grey panel, white bar, green speculum and dark-grey primaries. Breast Mottled brown and white. Abdomen Chestnut-brown with a white flank-spot. Bill Grey. Legs and feet Orange.

Description – female (breeding plumage) Upperparts and underparts Brownish. Upperwing Bluish-grey panel, white bar, green speculum and dark-grey primaries. Bill Grey.

Description – eclipse Upperparts and underparts Drab brown, although male birds retain some blue-grey tonings to the head. White flank-spot is absent.

Conspicuous features Spoon-shaped bill ✎ In breeding plumage the male bird is brightly coloured ✎ White flank-spot.

Conspicuous characteristics Birds sit lower in the water than Grey Ducks or Mallards ✎ Appearance is of a short neck and big head ✎ Birds swim in a hunched position ✎ Tend to stay together when in a group of waterfowl ✎ Regularly seen swimming in pairs ✎ Fast-flying, similar to Grey Teal ✎ In-flight birds show bright wing markings ✎ Birds prefer still water where water insects and larvae can safely breed ✎ Seldom seen away from water, unlike Grey Ducks and Mallards. This is because the heavy pasture seeds and shoots that Grey Ducks and Mallards eat are not able to be handled by the fine-edged, sieve-like, spoon bills of the Shoveler.

Call Female, a 'cuck cuck cuck' similar to the Grey Teal. Male bird, a 'clonk' sound.

Nest A bowl lined with down in thick grass or rushes. Four cream eggs are laid.

Target localities Straka Refuge, Weranui Road, Waiwera ✎ Waimeha Lagoon, Waikanae.

Blue Duck

Conspicuous colour Blue-grey.
Habitat Confined to clear water, forested, mountain rivers in secluded areas away from human interference. Birds will extend territories into river stretches which are bordered by pasture for short periods. Some birds, probably unmated juveniles, occasionally turn up on lakes and ponds away from mountain rivers during winter.
Range In the North Island birds are confined to the forested rivers in the central North Island. South Island birds are confined to similar habitats on the western side of the Southern Alps and into eastern areas of Fiordland.
Size 530 mm (House Sparrow 145 mm, Mallard 580 mm).
Abundance Rare.
Status Endemic.

Family Anatidae **Species** *Hymenolaimus malacorhynchus*

Common names Blue Duck, Whio

Description Upperparts Grey. In good light a bluish tinge may be evident. Birds are the colour of the mountain rocks around which they live. Breast Reddish spotting over grey. Abdomen and undertail Grey. Bill Off-white. Eye Iris, yellow. Legs and feet Brown.

Conspicuous features Larger than a Shoveler / Pale bill stands out / Upper mandible flaps on tip of bill make it different from other ducks / Yellow eye is noticeable.

Conspicuous characteristics Birds are always in pairs when on river territories. Single birds do turn up on ponds and lakes away from normal breeding habitats / Swimming birds often hold tail up / When feeding, birds are regularly seen with tails up and heads under the water. They also stand and dabble / Birds are quiet by nature and prefer to swim away from disturbance rather than fly / Birds tend to venture out and feed in early morning and late afternoon. By day they hide among overhanging river vegetation. However, daytime sightings are not uncommon.

Call Male bird a loud whistle – 'whio' (wh is f in Maori). Female birds answer with a harsh 'craak craak'. Usually they call at dusk or dawn.

Nest A bowl of grass lined with down, under a log or in dense vegetation. Up to six cream eggs are laid. Blue Ducks are strongly territorial, living on set stretches of river in territories of about 1 km of river long.

Note The Blue Duck is a unique New Zealand endemic river duck with special physical adaptations such as its bill shape, its feeding habits and its feet size. These specialised features allow it to survive in its mountain river environment. Here it feeds on a soft diet of river insect larvae and under-stone algae. Its bill, which is soft and pliable, with upper mandible flaps which tend to protect the lower mandible, is adapted for the taking of food among rough rocks. Its large feet enable it to swim strongly in turbulent waters from birth.

Target localities Lake Waikaremoana near the park headquarters / Waihaha River near Tihoi / Raetihi on the Maunganui Ateao River / Waioeka Gorge, Bay of Plenty / Arthurs Pass at Pegleg Creek / Eglinton Valley on Wesney Creek just before Knobs Flat.

Cape Barren Goose

Conspicuous colour Grey.
Habitat Parks, lakes and pasture.
Range Regularly sighted in Canterbury. Odd ones turn up away from waterfowl collections.
Size 870 mm (House Sparrow 145 mm, Feral Goose 800 mm).
Abundance Rare.
Status Introduced.

Family Anatidae **Species** *Cereopsis novaehollandiae*

Common name Cape Barren Goose

Description Upper and underparts Pale grey. Wings Pale grey with occasional black spots. Primary feathers Black. Legs Pink. Feet Black. Bill Greenish-yellow.

Conspicuous features Pale grey colouring Greenish-yellow bill.

Conspicuous characteristic Continually grazes with head down.

Call Usually silent but a high-pitched trumpet call at times.

Nest A bowl of grass and rushes on the ground. Five white eggs are laid.

Target localities Wetlands in the Canterbury district.

Grey Ternlet

Conspicuous colours Blue and grey.
Habitat Coast and offshore islands.
Range Waters to the east of the North Island.
Size 280 mm (House Sparrow 145 mm, White-fronted Tern 400 mm).
Abundance Uncommon.
Status Native.

Family Laridae **Species** *Procelsterna cerulea*

Common names Grey Ternlet, Grey Noddy

Description Crown Whitish with dark eye. Upperparts Delicate blue-grey with dark primaries on wings. Wings Blue-grey with a thin, white trailing edge. Underparts Light grey.

Conspicuous features Delicate blue-grey colouring ✎ Dark eye.

Conspicuous characteristics Perches on cliff ledges ✎ Flies out to fish and then returns to same ledge.

Note A bird of the tropics and subtropics which breeds on the Kermadec Islands, on West Island in the Three Kings group and Volkner Rocks off White Island.

Call Quiet purring 'croror or or', sounds.

Nest One cream, brown-blotched coloured egg is laid on a bare rock shelf.

Target locality Mokohinau Islands in waters out from Sandspit in summer.

Reef Heron

Conspicuous colours Blue and grey.
Habitat Coastal but sometimes found inland on the edges of lakes and wetlands which are not too distant from the headwaters of estuaries or harbours.
Range Around the coastline of both the North and South Islands and on Stewart Island. More often seen along the east coast of Northland than anywhere else.
Size 660 mm (House Sparrow 145 mm, White-faced Heron 670 mm).
Abundance Uncommon.
Status Native.

Family Ardeidae

Species *Egretta sacra*

Common names Reef Heron, Blue Heron, Matuku-moana

Phases In Australia and the Pacific Islands it comes in both white and grey phases with some blotchy intermediate types. New Zealand has only the grey phase. The Reef Heron of Africa (*E. gularis*), which comes in both a white and grey phase, is similar.

Description Upperparts and underparts Charcoal-grey but with a blue look. Bill Yellowish-brown. Legs Yellowish-green. Eye Iris is yellow.

Conspicuous features Birds are a darker grey than the more common White-faced Heron.

Conspicuous characteristics Hunched gait / Slightly smaller and heavier looking than the White-faced Heron / When feeding often raises its wings to shade the water to enhance vision.

Call A guttural croaking often indicates a bird about to land.

Nest A solitary nester, making a nest of twigs and dried seaweed on the ground in a cave or rock crevices. Up to four pale blue-green eggs are laid.

Target localities Waitangi Estuary, just north of Paihia / Aroha Island, Kerikeri / Weiti River (Wade River), Whangaparaoa Peninsula / Raglan Harbour, west of Hamilton / Motueka Estuary, Marlborough.

White-faced Heron

Conspicuous colours Blue and grey.
Habitat Both coastal and inland on the edges of lakes and wetlands.
Range Throughout New Zealand.
Size 670 mm (House Sparrow 145 mm, Reef Heron 660 mm).
Abundance Common.
Status Native. Self introduced from Australia and first confirmed breeding in 1941.

Family Ardeidae

Species *Ardea novaehollandiae*

Common name White-faced Heron

Description — mature
Forehead, face and chin White.
Upperparts Bluish-grey. Long, pale-grey plumes on back can be seen in the breeding season.
Wings Bluish-grey with dark-grey primaries.
Underparts Light grey.
Bill Black.
Legs and feet Greenish-yellow.

Description — immature
Face Bluish-grey with a white chin. Juveniles lack the white face but retain a small amount of white on the chin. Wings Bluish-grey, with black primaries both under and on top of the wings.

Conspicuous features Face and forehead distinctly white ✎ Plumage is more bluish than that of the Reef Heron.

Conspicuous characteristics Has a less hunched gait than the Reef Heron and appears slightly larger ✎ Often seen sitting on the edges of farm water troughs ✎ Often seen feeding on wet pasture ✎ Nests high in old conifers or eucalypt trees ✎ Flight is slow and wafting. Head is usually tucked in a hunched position ✎ Will fly with neck extended.

Call Makes a guttural croaking sound, especially when approaching the nest.

Nest A bundle of sticks high in a tree. Up to four pale blue-green eggs are laid.

Target localities Kawakawa Bay, east of Papakura ✎ Miranda, Firth of Thames ✎ All Day Bay, south of Oamaru ✎ Open country in most districts.

Spotted Shag

Conspicuous colours Blue and grey.
Habitat Coast and rock stacks.
Range In the North Island most of the population is confined to the Hauraki Gulf and around the Wellington coast. A few breed on the cliffs at Bethells Beach (Te Henga), west of Auckland. In the South Island the range is from Marlborough to Otago with the Blue Shag (*S. p. steadi*) being found around eastern Southland and up the West Coast. It is absent from Fiordland.
Abundance Common.
Size 700 mm (House Sparrow 145 mm, Mallard 580 mm).
Status Endemic.

Family Phalacrocoracidae

Species *Stictocarbo punctatus*

Common names Spotted Shag, Parekareka

Subspecies These are *S. p. punctatus*, found in the north and around the northern parts of the South Island, and the Blue Shag (*S. p. steadi*), found in coastal Southland, Stewart Island and parts of coastal Westland.

Description – *S. p. punctatus* Head Black, with a distinctive white head and neck stripe starting from above the eye and down the neck. Naked skin around eye Greenish-blue. Crest Black in breeding plumage birds. Wings Greyish-brown with black spotting. Breast Soft grey. Underwings and undertail Black. Bill Light brown. Legs and feet Orange.

Description – *S. p. steadi* Upper and underparts Similar to *S. P. punctatus* but white neck strip slimmer, more blue tonings on breast, and darker wings with less spotting.

Conspicuous features Black head crest on breeding birds ✎ White head and neck stripe ✎ Blue-grey breast.

Conspicuous characteristics Roost on coastal rocks with the incoming tide ✎ Regularly feed up to 15 km out to sea ✎ Often congregate in giant feeding rafts on the water, by day, where food is plentiful ✎ Congregate at breeding-colony roosts at night and can be seen in late afternoon flying low over the water as they head towards them ✎ Hold wings out to dry ✎ Fly low with head and neck straight out.

Call Usually silent but grunts and croaks can be heard when courtship displaying.

Nest A colonial breeder, nesting on cliffs or rocks. Nests are made of grass and seaweed. Up to three pale-blue eggs are laid.

Target localities Bethells Beach (Te Henga), west of Auckland ✎ Waiomu Beach, north of Thames ✎ Waikanae Estuary, north of Wellington ✎ Goose Bay, Oara, Kaikoura, North Canterbury ✎ Oamaru Harbour breakwater, near the Blue Penguin colony ✎ Taiaroa Head, Otago Peninsula ✎ Paterson Inlet, Stewart Island.

Differences between Spotted Shag and Blue Shag

- On the Spotted Shag the white stripe, which extends from above the eye and down the side of the neck, is wider.
- On the Spotted Shag the black spotting on greyish-brown wings is clearly visible.

Blue Penguin

Conspicuous colours Blue and grey.
Habitat Sea.
Range Around New Zealand and southern Australia.
Size 400 mm (House Sparrow 145 mm, Common Diving Petrel 200 mm).
Abundance Common.
Status Native.

Family Spheniscidae

Species *Eudyptula minor*

Common names Blue Penguin, Little Blue, Korora

Subspecies Six subspecies have been suggested based on localities. They are *E. m. novaehollandiae*, Australia and Tasmania; *E. m. iredalei*, North Island; *E. m. variabilis*, Cook Strait; *E. m. albosignata*, Banks Peninsula; *E. m. minor*, South Island and Stewart Island; E. m. chathamensis, Chatham Islands. Currently none have been recognised although most treat the White-flippered Penguin (*E. m. Albosignata*) as a subspecies, defined by the extensive white on its flippers.

Description Upperparts Blue. Throat Grey. Underparts White. Flippers Blue, edged with white. Note that the White-flippered subspecies has very bold white leading edges to the flippers. Bill Black.

Conspicuous features Small size when compared with other penguins ✎ Heavy bill ✎ Blue colouring.

Conspicuous characteristics When on the sea surface only the head and the upperpart of the back are noticeable ✎ At times birds loaf together on the water in rafts ✎ Birds 'porpoise' when swimming towards the shore.

Call At sea a duck-like quack is often heard. At the nesting site, deep-toned growls with some trumpeting and mewing sounds.

Nest Breeding season is from August to February. Nests in burrows or rock cavities, sometimes high on cliff faces. Burrows are usually isolated but within calling distance of each other. Birds return to the burrows just after dark, hopping from the sea on to rocks.

The moult After the breeding season birds come ashore and hide in caves for periods of up to two weeks and moult.

Target localities Waters out from Sandspit ✎ Waters of the Tiritiri Matangi channel ✎ Oamaru Harbour at the breakwater at the end of Waterfront Road ✎ Waters of Milford Sound ✎ Waters around Stewart Island and in Paterson Inlet.

White-flippered Penguin

Conspicuous colours Blue and grey.
Habitat Sea.
Range Found around Banks Peninsula near Christchurch in the South Island.
Size 400 mm (House Sparrow 145 mm, Common Diving Petrel 200 mm).
Abundance Common.
Status Native.

Family Spheniscidae

Species *Eudyptula minor albosignata* – considered a subspecies of *E. minor*

Common name White-flippered Penguin

Description Flippers The main difference, when compared with the Blue Penguin, is the heavy, white leading edge to flipper, and the whiter underparts.

Conspicuous feature Heavy, white leading edge to flippers.

Target locality Around Banks Peninsula and Akaroa Harbour.

Rifleman (female)

Conspicuous colour Brown.
Habitat Forests.
Range Found through central and lower North Island and the South Island excluding the Canterbury Plains and North Canterbury pasture. Also on Stewart Island.
Size 80 mm (House Sparrow 145 mm, Grey Warbler 100 mm).
Abundance Common in selected areas.
Status Endemic.

Family Acanthisittidae

Species *Acanthisitta chloris*

Common names Rifleman, Titipounamu

Subspecies Two subspecies are recognised: North Island Rifleman (*A. c. grantii*) and South Island Rifleman (*A. c. chloris*).

Description – female
Head Brown, slightly zebra-striped with white.
Eyebrow White.
Upperparts Brown, slightly zebra-striped with white.
Wings Brown, with faint yellow bars on the secondary feathers and faint white markings on the inner edges of primary feathers. Underparts Whitish. Tail Brown, and white-tipped.

Description – male Head Bright yellow-green with white eye-stripe. Upperparts Bright yellow-green. Wings Green, with a noticeable yellow bar. Underparts Whitish. Tail Brown, and white-tipped.

Conspicuous features The smallest bush bird ✎ Smaller than the Grey Warbler ✎ Appears tailless in comparison to the Grey Warbler ✎ White eyebrow on both male and female birds ✎ White tip to tail ✎ Pale-yellow wingbars.

Conspicuous characteristics Has a habit of working up tree trunks hopping from one side to the other ✎ Sometimes it wing-flicks as it feeds.

Call A high-pitched 'zit zit zit zit' that is difficult to hear.

Nest A hole or cavity in an old tree. Up to four white eggs are laid.

Target localities Pureora Forest, Benneydale, King Country ✎ Opepe Reserve, State Highway 5, Taupo ✎ Lake Rotopounamu, Tokaanu ✎ Lake Waikareiti track near Lake Waikaremoana ✎ Pelorus Bridge, Marlborough ✎ Lake Rotoiti, Marlborough ✎ Hinewai Reserve, Akaroa ✎ Lake Gunn, Fiordland.

Rock Wren (female)

Conspicuous colour Brown.
Habitat Alpine areas above the tree line, and in rock gardens of scattered low vegetation along either side of the Southern Alps.
Range South Island only. Populations are higher in the Fiordland area.
Size 100 mm (House Sparrow 145 mm, Grey Warbler 100 mm).
Abundance Uncommon in selected areas.
Status Endemic.

Family Acanthisittidae

Species *Xenicus gilviventris*

Common names Rock Wren, Hurupounamu

Description – female
Upperparts Olive-brown. Underparts Grey-brown. Flanks Yellow and green but paler than male.

Description – male
Head Dull green. Eyebrow White, edged with black. Upperparts Dull green. Underparts Grey-brown. Flanks Yellow and green. Bill Black and fine.

Conspicuous features Larger than the Rifleman ✎ White eyebrow ✎ Yellow and green on flanks under primaries ✎ Large toes, with prominent hind-toe.

Conspicuous characteristics Continually bob their bodies when standing ✎ Will suddenly appear ✎ Hop quickly from rock to rock with some use of wings ✎ At times they undertake longer flights ✎ Have the appearance when in flight of young sparrows trying their wings ✎ Feed on moths and insects among the rocks ✎ A rare species, difficult to find.

Call A high-pitched, three-syllable 'tzee tzit tzit' sound, the first note being rather piercing. The call is not often heard but is distinctive.

Nest A hole in a bank or rock crevice on the ground. Up to three creamy eggs are laid.

Target localities Arthur's Pass National Park ✎ Gertrude Valley, Hollyford Valley, Fiordland ✎ Homer Tunnel, Milford road, Fiordland ✎ Start of the Routeburn Track, Glenorchy, near Queenstown.

Tomtit (female)

Conspicuous colour Brown.
Habitat A bird of old forests, secondary forest and exotic pine forests.
Range Found throughout New Zealand and on offshore islands.
Size 130 mm (House Sparrow 145 mm, Grey Warbler 100 mm).
Abundance Common.
Status Endemic.

Family Eopsaltriidae

Species *Petroica macrocephela*

Common names Tomtit, Miromiro (North Island), Ngirungiru (South Island)

Subspecies Five: North Island (*P. m. toitoi*), South Island (*P. m. macrocephala*), Chatham Island (*P. m. chathamensis*), Snares Island (*P. m. dannefaerdi*), Auckland Island (*P. m. marrineri*).

Description – female Upperparts Brownish. Frontal dot above bill Less conspicuous than on male bird and sometimes absent. Wingbar White, but less obvious than on male birds. Chin and breast Grey-brown. Underparts Greyish.

Description – male Head, throat and upperparts Black. Frontal dot above bill White. Wings Black with a conspicuous white wingbar. Underparts Pure white on the North Island bird and white washed with yellow on the South Island, Stewart Island and Chatham Island birds. Tail Black with white edges.

Conspicuous features Bigger and heavier than the Grey Warbler ✎ Has prominent eye.

Conspicuous characteristics Of confiding nature ✎ Birds often are seen clinging to the side of trunks with head down.

Call Song, delivered by the male bird, is a fragile 'sweedle sweedle sweedle swee'. Females occasionally sing a weaker version of the male song.

Nest Of twigs and bark, lined with moss in a cavity. Four cream-coloured eggs are laid.

Target localities Dome Valley walk, Warkworth ✎ Pine forests near Rotorua ✎ Eglinton Valley near Te Anau.

Brown Creeper

Conspicuous colour Brown.
Habitat Found in any type of forest and even small remnants from mid-storey to canopy. Occasionally in scrubland and roadside verges.
Range Found over much of the South Island but not on the Canterbury Plains. Also on Stewart Island and its outlying islands, including Codfish Island. Absent from the North Island.
Size 130 mm (House Sparrow 145 mm).
Abundance Common.
Status Endemic.

Family Pachycephalidae

Species *Mohoua novaeseelandiae*

Common names Brown Creeper, Pipipi

Description Crown Bluish-grey colouring to below eye. Face and neck Ash-grey with white eyebrow behind eye. Upperparts Cinnamon brown. Underparts Buff brown.

Conspicuous features Distinct line below the eye separates the bluish-grey head from the buff underparts. The eye is difficult to see.

Conspicuous characteristics Outside of the breeding season birds congregate in large, noisy flocks and move through the canopy in this manner. Can move secretively through low vegetation.

Call A rapid 'ddee did dee dee', of fragile quality. Song not unlike that of the Yellowhead.

Nest A deep cup of twigs and leaves lined with moss or lichen. Three white, brown-speckled eggs are laid.

Target localities Forests around Picton on verges of the Marlborough Sounds. Forests of the Eglinton Valley, Te Anau.

Fantail (non-breeding)

Conspicuous colour Brown.
Habitat Found anywhere where there is vegetation and in suburban gardens.
Range Throughout New Zealand and the offshore islands.
Size 160 mm (House Sparrow 145 mm, Grey Warbler 100 mm).
Abundance Common.
Status Native.

Family Monarchidae

Species *Rhipidura fuliginosa*

Common names Fantail (There are several common Maori names, varying from one part of the country to the other, including Piwakawaka, Piwaiwaka, Tiwakawaka, Tiwaiwaka, Tirairaka.)

Subspecies North Island Fantail (*R. f. placabilis*), South Island Fantail (*R. f. fuliginosa*), Chatham Island Fantail (*R. f. penita*).

Description – pied phase (non-breeding) Upperparts Dull brown. Underparts Fawn, lacking yellowish wash. Chin Faint white line. Tail Two central feathers, black. Outer feathers, white.

Description – pied phase (breeding) Head Brownish-black with a white eyebrow. Upperparts Brown. Chin White, below which is a black bar. Underparts Tan colouring but yellowish when in full breeding plumage. Tail As for non-breeding bird.

Conspicuous features Overall drab look White, striped with black, fan-like tail.

Conspicuous characteristics During late summer non-breeding birds reappear Will enter houses and sit on lampshades Always active, constantly shifting position with sideways movement of tail.

Call A high-pitched 'cheet cheet' communication call. When in nesting territories male birds have a vocal and constant, repetitive chattering call.

Nest Neat wine glass-shaped cup of twigs and leaves lined with moss. Four white, brown-speckled eggs are laid.

Target localities Found everywhere from gardens to forest.

Stitchbird (female)

Conspicuous colour Brown.
Habitat A bird of both secondary and mature native forests.
Range Little Barrier Island, Tiritiri Island and Kapiti Island.
Size 180 mm (House Sparrow 145 mm, Bellbird 200 mm).
Abundance Rare.
Status Endemic.

Family Meliphagidae

Species *Notiomystis cincta*

Common names Stitchbird, Hihi

Description – female Upper and underparts Olive-brown, similar to the female Bellbird. Wings Brown with a white wingbar.

Conspicuous features White wingbar separates the female Stitchbird from the female Bellbird ✎ Has whiskers at the gape.

Conspicuous characteristics Birds are usually in pairs ✎ Strongly territorial in the nesting season ✎ Out of the nesting season often follow flocks of Whiteheads through the forest ✎ Have a habit of alighting on a branch with the tail held high or even over their backs or with heads facing down.

Call A 'pek pek pek' alarm call similar to the female Bellbird.

Nest Usually in holes where a platform of sticks in built. Four white eggs are laid.

Target locality Tiritiri Matangi Island.

Bellbird (female)

Conspicuous colour Brown.
Habitat Old forests, secondary forests, scrublands and in some areas in suburban gardens.
Range Well-spread throughout the forested areas of the South Island and forested areas of the North Island north to the Waikato and to the top of the Coromandel Peninsula. Absent from Northland, although establishing on Whangaparaoa Peninsula. Also on offshore islands but not Great Barrier Island.
Size 200 mm (House Sparrow 145 mm, Tui 300 mm).
Abundance Common in selected areas.
Status Endemic.

Family Meliphagidae

Species *Anthornis melanura*

Common names Bellbird, Korimako, Makomako

Subspecies Three are recognised: Bellbird (*A. m. melanura*), Three Kings Bellbird (*A. m. obscura*), Poor Knights Bellbird (*A. m. oneho*).

Description – female Upperparts Olive-brown. Cheek Whitish stripe under eye from gape. Wings and tail Brownish-black, tail slightly forked. Underparts Pale-green. Bill Black.

Description – male Head Olive-green with purple iridescence on forehead and crown. Upperparts Olive-green. Wings Dark bluish-black with yellow at bend of folded wing. Tail Bluish-black and slightly forked. Underparts Pale-green. Eye Red. Bill Black.

Conspicuous features Curved bill • Slightly forked tail.

Conspicuous characteristics Early-morning chorus singer • Takes nectar from flowers in an acrobatic manner • Flight is fast with noisy wing rustle • Flight is manoeuvrable in forests • In the open its flight is direct with some undulation.

Call Song is bell-like, liquid, clear and melodic and starts before sunrise. Both sexes sing, male birds being stronger and repetitive. The female song is a series of individual notes. Both make 'pek pek pek' alarm notes.

Nest A loosely built structure of twigs and fern leaves lined with grasses 3–4 m above ground. Four white, brown-blotched eggs are laid.

Target localities Tiritiri Matangi Island • Kauaeranga Valley, Thames • Mt Peel Forest, Canterbury • Stewart Island.

Morepork

Conspicuous colour Brown.
Habitat A bird of old forests, modified forests, exotic forests and even well-vegetated suburban gardens. Most of the big cities have breeding populations in their wooded suburbs but it is not found in Christchurch.
Range North Island, South Island and Stewart Island but it is absent from pastoral areas of limited vegetation.
Size 290 mm (House Sparrow 145 mm, Little Owl 230 mm).
Abundance Common.
Status Native.

Family Strigidae

Species *Ninox novaeseelandiae*

Common names Morepork, Ruru

Description Head Brown, sometimes with a white ring around the crown and with whitish eyebrows. Upperparts and underparts Mostly brown. Feathers are lightly flecked with white, tan or creamy colouring, especially around eyes and breast. Eyes Yellow, set in a dark face. Legs Brownish-yellow, with brown feathers and brown toes.

Conspicuous features and characteristics

Staring eyes placed on the front of the face ✎ Nocturnal in feeding and calling habits ✎ Silent in flight ✎ Will sit on power wires at night near a street light and catch light-attracted insects ✎ Favours the same daylight-roosting branch from day to day.

Call A monotonous 'more-pork more-pork' (probably male bird only). Also 'cree cree' screeches are made and 'more more more' sounds.

Nest A heap of sticks in a hole or old tree trunk. Up to three white eggs are laid.

Target localities Cornwall Park in Auckland ✎ Botanic Garden in Wellington.

Little Spotted Kiwi

Conspicuous colour Brown.
Habitat Forest and low bush.
Range The largest population of about 1000 birds is on Kapiti Island. Also on Hen Island, Tiritiri Matangi Island, Red Mercury Island and Motuara Island in Queen Charlotte Sound. Recently released in selected 'mainland island' sites.
Size 300 mm (House Sparrow 145 mm, Brown Kiwi 500 mm).
Abundance Rare.
Status Endemic.

Family Apterygidae **Species** *Apteryx owenii*

Common names Little Spotted Kiwi, Kiwi-pukupuku

Description Upper and underparts Brownish grey, finely mottled and banded with white. Bill Grey.

Conspicuous features Smaller than the Brown Kiwi / Bill and body proportions similar to Brown Kiwi / Density of white mottling finer than on Great Spotted Kiwi.

Call Similar to Brown Kiwi but of faster delivery and higher note.

Nest Hole in a log or in the ground. Similar to Brown Kiwi.

Target localities Tiritiri Matangi Island / Kapiti Island.

Brown Kiwi

Conspicuous colour Brown.
Habitat Native and exotic forests and scrublands.
Range Northland, central North Island, Fiordland and Stewart Island. Also on some offshore islands including Moturoa Island (Bay of Islands), Kawau Island, Little Barrier Island and Ponui Island.
Size 400 mm (House Sparrow 145 mm, domestic hen 500 mm).
Abundance Uncommon.
Status Endemic.

Family Apterygidae **Species** *Apteryx australis*

Common names Brown Kiwi, North Island Brown Kiwi

Species and subspecies There are five formally recognised species of kiwi: Brown Kiwi (*Apteryx mantelli*) in the North Island, Rowi or Okarito Brown Kiwi (*A. rowi*) at Okarito, Tokoeka (*A. australis*) in Haast, Fiordland and on Stewart Island, Great Spotted Kiwi (*A. haastii*) and Little Spotted Kiwi (*A. owenii*). Kiwi are New Zealand's most unique and ancient birds. The fact that they are flightless and have only rudimentary wings and no tail suggests arrival here prior to the separation of the ancient landmass of Gondwanaland. They are members of the ratite grouping of birds, which includes the extinct New Zealand moa species, and the Ostrich, Emu, Cassowary and Rhea.

Description Upperparts Brown, streaked with dark brown. Feathers are hair-like, being both shaftless and barbless. Underparts Similar colouring to upperparts. Bill Grey.

Conspicuous features Can't fly ✎ Has long bill with nostrils at the end ✎ Has no tail.

Conspicuous characteristics Nocturnal ✎ When feeding it often sits on its haunches. In this position it can look like a mound of brown dirt or even a hedgehog ✎ When running away, it has a waddling, side to side, swaying gait ✎ Feeding Brown Kiwi regularly make heavy breathing-like, snuffling sounds.

Calls Starts calling about 40 minutes after sundown and continues intermittently throughout the night with a noisy period before dawn. The call of the male bird is a shrill 'ah-el, ah-el', uttered several times. The female has a quieter and hoarser call 'aarh, aarh aarh' or 'ah-eh, ah-eh'. (Description: Colbourne 1981.)

Nest and breeding season Female Brown Kiwi are unique in the size of the egg they lay, which can be equivalent to a quarter of the female's body weight. Usually two white eggs are laid in burrows, cavities, or under vegetation and trees. These can be incubated by both sexes but more often, especially with North Island birds, by the male, which will also care for the young within the burrow in the early stages. Prior to nesting, male birds are known to be strongly territorial. Birds are thought to breed throughout the year with the July to February months being the most common.

Most vocal months Late winter which indicates mating and the start of breeding.

Signs that show the presence of Brown Kiwi Calls and snuffling sounds and sometimes the sounds of breaking vegetation, caused as they push through the undergrowth ✎ Footprints in mud and on tracks. Brown Kiwi footprints are bold and bigger than those of a domestic hen. Usually only three toes are showing, unless the bird has been walking through deep and soft sand when the hind toe might be seen ✎ White, pungent-smelling dung on trails or around forest burrows ✎ Feeding probe marks in the ground, usually with a circular whirl at the entrance.

Target localities Aroha Island, Kerikeri, a small island linked to the mainland by a causeway ✎ Waitangi Forest, owned by Rayonier NZ Ltd, not far from the coastal town of Paihia ✎ Stewart Island, where there is a commercial tourism operator.

Great Spotted Kiwi

Conspicuous colour Brown.
Habitat Forest.
Range Found down the west coast of the South Island to Franz Josef Glacier, with biggest populations in the Paparoa Range south of Westport, where they are sustaining themselves. Also recorded in Arthur's Pass National Park.
Size 450 mm (House Sparrow 145 mm, Brown Kiwi 500 mm).
Abundance Uncommon.
Status Endemic.

Family Apterygidae **Species** *Apteryx haastii*

Common names Great Spotted Kiwi, Roa

Description Upper and underparts Brownish-grey, finely banded with white. Bill Grey.

Conspicuous features Slightly larger than the Brown Kiwi ✎ Bill and body proportions similar to Brown Kiwi.

Call Higher pitched than Brown Kiwi but similar.

Nest Similar to Brown Kiwi but only one white egg is laid.

Target localities Paparoa Range between Westport and Greymouth ✎ Lake Brunner, east of Greymouth ✎ Rough Creek near Arthur's Pass.

New Zealand Falcon

Conspicuous colour Brown.
Habitat The Bush Falcon and Southern Falcon are confined to forested areas while the Eastern Falcon is widely spread over pasture and rough and open high country.
Range In the North Island widespread south of a line from Kawhia Harbour to the central Bay of Plenty. In the South Island it is widespread throughout but nowhere common. Also on Stewart Island and Auckland Islands.
Size 430 mm (House Sparrow 145 mm, Harrier 550 mm).
Abundance Uncommon.
Status Endemic.

Family Falconidae

Species *Falco novaeseelandiae*

Common names New Zealand Falcon, Bush Falcon, Sparrowhawk, Karearea

Subspecies Three: Bush Falcon of North Island and northwest of the South Island ✎ Eastern Falcon of the eastern South Island ✎ Southern Falcon of Fiordland and Auckland Islands.

Description – Bush Falcon Crown, nape and back Bluish-black. Wings Bluish-black, with grey barring and pure black primary feathers. Underwings Brown, barred with white. Uppertail Black, barred with separate narrow white bars, 7–11 in number. Undertail Buff, with some grey barring. Throat and breast Cream, vertically striped with dark brown. Flanks Cream, horizontally barred with dark brown. Ceres (above bill) Yellow. Eye Iris dark brown with a thin, yellow eye-ring. Bill Black, with dark brown markings on side of bill and chin. Legs and feet Yellow.

Description – Eastern Falcon Paler than the Bush Falcon and the Southern Falcon and more richly barred.

Description – Southern Falcon Paler than the Bush Falcon with more under-barring on male birds.

Description – immature Upper and underparts Brown. Chin Cream. Ceres Grey.

Conspicuous features Yellow ceres and yellow legs ✎ Dark-brown eye and yellow eye-ring separates it from the Australasian Harrier ✎ Black claws and beak.

Conspicuous characteristics When flying low and straight, with tail out, it can be mistaken for a New Zealand Pigeon or a Long-tailed Cuckoo ✎ In fine weather birds will rise and spiral on thermals ✎ Hovers with quick wing-beats ✎ Will fly out in a slow, hovering-type flight and investigate human visitors to a territory.

Call A rapid, high-pitched, shrieking 'kek kek kek'.

Nest The Bush Falcon and Southern Falcon nest high in trees, often in astelia. The Eastern Falcon nests in a scrape on a high rock ledge. Three reddish-brown, dark brown-blotched eggs are laid.

Target localities Pureora Forest Park, Benneydale ✎ Lake Gunn, Eglinton Valley, near Te Anau ✎ Lake Manapouri township.

Weka

Conspicuous colour Brown.
Habitat Forests, scrubland, fringes of estuaries, coastal beaches, roadside verges and rough pasture.
Range In the North Island, was confined to isolated areas such as Rakitu (Arid) Island, Kawau Island, Mokoia Island, Raukumara Forest Park, Tolaga Bay, and Kapiti Island, but recently populations can be found at Russell, Bay of Islands, and Kawakawa Bay, east of Auckland. In the South Island the bulk of the population is in the northwest corner, Westland and Fiordland. Also on Stewart Island.
Size 530 mm (House Sparrow 145 mm, Banded Rail 300 mm).
Abundance Common in selected areas.
Status Endemic.

Family Rallidae

Species *Gallirallus australis*

Common name Weka

Subspecies Has four distinct subspecies, based on plumage colour patterns: The North Island race (*G. a. greyi*) – greyish underparts and brown legs The eastern South Island race (*G. a. hectori*), known as the Buff Weka – sandy-brown colouring and reddish legs The western South Island race, known as Western Weka or Black Weka (*G. a. australis*) – dark blackish-brown colouration, although some birds lack the blackish colouring and appear instead deep chestnut-brown The Stewart Island Weka (*G. a. scotti*) – paler colourings all over compared with Western.

Description Upperparts and underparts Brown with black markings. (See above under Subspecies for variations.)

Conspicuous features Like a slim version of a domestic fowl but with a heavier bill.

Conspicuous characteristics A flightless bird, capable of very fast running Slinks through the forest with head low to the ground and tail raised Inquisitive, and will warily approach humans Defends its territories with fast chases North Island birds are less confiding than South and Stewart Island birds.

Call A loud, rapidly repeated 'cooeet cooeet cooeet', heard day or night.

Nest A bowl of grass in rushes or tall grass. Up to five pinkish eggs are laid.

Target localities Kawau Island, accessed from Sandspit Mokoia Island, Lake Rotorua Pelorus Bridge west of Picton Stewart Island.

Redpoll

Conspicuous colour Brown.
Habitat A bird of open country, wasteland and native and exotic forests.
Range Found throughout New Zealand but more common in the South Island.
Size 120 mm (House Sparrow 145 mm, Chaffinch 150 mm).
Abundance Common.
Status Introduced.

Family Fringillidae

Species *Carduelis flammea*

Common name Redpoll

Description – male

Head Dark brown behind bill and then a reddish crown to the back of the head.
Neck, nape and back Brownish.
Wings Dark brown on primaries with distinctive whitish wingbar.
Chin Black.
Breast Brownish with a reddish flush in the breeding season.
Underparts Greyish with some scattered dark-brown striping.

Description – female Crown Reddish. Upperparts Greyish-brown with noticeable dark-brown striping, but less coloured than on the male bird. Underparts Whitish with light-brown striping but lacking the reddish flush of breeding males.

Conspicuous features Whitish wingbar ✎ Red on forehead ✎ Black chin on male birds ✎ Streaking on the back of head and mantle.

Conspicuous characteristics Has a noisy flight song ✎ Feeds on the ground, often among other finches ✎ Will feed close to buildings and houses in some localities ✎ Enjoys feeding over stubble paddocks.

Call Song, which is delivered by the male bird, is a loud, direct, rippling trill, more often heard in flight than from a perch.

Nest Small, neat cup of grass lined with wool. Four bluish-green eggs are laid.

Target localities Countryside in South Waikato districts ✎ Queenstown district ✎ Oban, Stewart Island.

Hedge Sparrow

Conspicuous colour Brown.
Habitat Low vegetation and second growth. Also pine forests, well-established suburban gardens and verges of native forest areas.
Range Found throughout New Zealand.
Size 140 mm (House Sparrow 145 mm).
Abundance Common.
Status Introduced.

Family Prunellidae

Species
Prunella modularis

Common names
Hedge Sparrow, Dunnock

Description
Head and upperparts Grey streaked with brown.
Face, throat, collar and underparts Soft grey.
Wings Brown streaked with dark brown.

Conspicuous features
Has fine bill as opposed to a House Sparrow's blunt bill ✎ Appears similar to a female House Sparrow but finer and smaller.

Conspicuous characteristics Sings in conspicuous places, often from in a roadside verge ✎ Feeds low down in shrubs and often on the ground ✎ Will venture out on to lawns sometimes in the company of House Sparrows ✎ A neat-looking bird of sombre colouring.

Call Its song, uttered by the male bird, is a fragile warble of 'sweedle sweedle swee swee' notes. It is the first of the introduced songsters to commence singing in the autumn and birds are in full song by mid-April in the north.

Nest A neat cup made of twigs and leaves, close to the ground and well hidden. Four blue eggs are laid.

Target localities Roadside verges in Waikato farming districts ✎ Wellington suburbs ✎ Oban, Stewart Island.

House Sparrow (female)

Conspicuous colour Brown.
Habitat In towns and suburbs and in rough open countryside.
Range Found throughout New Zealand and on some of the outlying islands.
Size 145 mm (Chaffinch 150 mm).
Abundance Common.
Status Introduced.

Family Ploceidae

Species *Passer domesticus*

Common name Sparrow

Description – female
Upperparts Brown with buffy patches to sides of face and neck. Wings Brown streaked black with a white wingbar but paler than male bird. Underparts Grey.

Description – male
Crown Grey. Upperparts Brown. Wings Brown, streaked black with a white wingbar. Rump Greyish-brown. Underparts Greyish-white.

Conspicuous feature Whitish wingbar.

Conspicuous characteristics Tends to flock in large numbers on pasture and roadside verges Returns to noisy evening roosts outside of the breeding season A late nester, breeding from early October when it can be seen carrying streamers of straw into the treetops or under house eaves Can be confused with other finches, especially the Redpoll and also the Hedge Sparrow.

Call A noisy chirrup and chattering sound is the usual song. At dawn and dusk House Sparrow evening roosts become very noisy. The dawn chorus starts after the Blackbird and Song Thrush have finished singing.

Nest A bundle of straw in trees or buildings with an entrance hole. Four white, brown-spotted and streaked eggs are laid.

Target localities Common everywhere.

Chaffinch (female)

Conspicuous colour Brown.
Habitat Suburban gardens, parkland, farmland, scrublands. Pine forests and native forests.
Range Found throughout New Zealand and on offshore islands.
Size 150 mm (House Sparrow 145 mm).
Abundance Common.
Status Introduced.

Family Fringillidae

Species *Fringilla coelebs*

Common name Chaffinch

Description – female Upperparts Brownish-grey. Wings Dark-brown primaries marked with two white wingbars. Rump Green. Underparts Greyish-brown. Tail Light brown in centre with outer feathers darker and edged with white.

Description – male Forehead Black. Crown and nape Bluish-grey. Face, throat and underparts Pinkish-brown. Back Reddish-brown. Rump Olive. Wings Black with two conspicuous white wingbars on each wing. Tail Central feathers brown. Outer feathers black with white outer edges.

Conspicuous features Greenish rump ✐ White wingbars.

Conspicuous characteristics Feeds on the ground but flies up for cover when disturbed ✐ Shows a lot of white on wings and tail when in flight ✐ In flight, when in forests, it is a faster and a more direct flier than native birds ✐ Has a habit of hawking insects like a flycatcher, especially over water ✐ During winter and early spring joins mixed flocks of other finches and feeds on lawns and pasture.

Call Not vocal like the male. Utters metallic 'pink pink pink' sounds.

Nest Neat cup made of twigs camouflaged with lichen. Four greyish-blue, purple-blotched eggs are laid.

Target localities Found everywhere.

Cirl Bunting (female)

Conspicuous colour Brown.
Habitat A bird of open pasture and low vegetation.
Range Common in northern South Island. Otherwise only sparse records.
Size 160 mm (House Sparrow 145 mm, Yellowhammer 160 mm).
Abundance Common in selected localities.
Status Introduced.

Family Emberizidae

Species *Emberiza cirlus*

Common name Cirl Bunting

Description – female Upperparts and underparts Dull yellow, streaked heavily with dark brown on the crown and light brown elsewhere. Rump Olive-brown.

Description – male Crown and nape Dark grey, striped with black and yellow. Face Yellow, with bold black stripe through eye. Chin and throat Black, with yellow band below. Neck and upper breast Pale brown. Wings Brown, edged with dark brown. Underparts Pale yellow, lightly streaked with brown. Rump Olive-brown. Tail Olive-brown, with black outer feathers, edged white.

Conspicuous features Yellowhammer size ✎ At first sight similar to a brownish Yellowhammer.

Conspicuous characteristics Feeds on the ground with Yellowhammers, especially among fed-out hay ✎ Hops while feeding on ground ✎ Often crouches close to ground like a Yellowhammer ✎ Twitters in flight when in family parties.

Call Female song is a rather thin 'zitt zitt' sound.

Nest A cup of grass, well hidden near the ground. Three bluish-green, brown-speckled eggs are laid.

Target localities Areas of pasture near Motueka, Marlborough district ✎ Oamaru racecourse, north of Oamaru.

Differences between Cirl Buntings and Yellowhammers

✎ *Female:* Cirl Buntings have olive-brown rumps; Yellowhammers have rufous rumps. ✎ Cirl Buntings are darker with less yellow.

✎ *Male:* Cirl Buntings have greyish crowns; Yellowhammers have yellow crowns. ✎ Cirl Buntings have black chins, and yellow faces separated by a bold black eye-stripe.

Skylark

Conspicuous colour Brown.
Habitat Open pasture and wilderness grassland areas.
Range Throughout New Zealand away from forests and alpine areas.
Size 180 mm (House Sparrow 145 mm).
Abundance Common.
Status Introduced.

Family Alaudidae **Species** *Alauda arvensis* **Common name** Skylark

Description Head Brown streaked with light brown. A crest is sometimes noticeable. Upperparts Brown streaked with light brown. Underparts Buff. Tail Dark brown, the outer feathers being white.

Conspicuous features Crest on head (not always visible) ✎ White tail feathers.

Conspicuous characteristics Sings purposefully from the sky ✎ Crouches on gravel country roads, flying off just as traffic approaches ✎ Occasionally sits on fence posts ✎ Flies up in front of intruders when in long grass ✎ Raises its crest when carrying out courtship and mating procedures.

Call Song, delivered by the male bird, is a liquid trill uttered when flying above nesting territories. Sometimes, though, it is delivered from atop a fence post or even when sitting on the ground. Birds can remain singing in the air for some time. Birds disturbed on the ground utter a 'chirrup' call.

Nest A cup of grass on the ground. Three cream-coloured, brown-speckled eggs are laid.

Target localities Any grassland or open countryside ✎ Miranda, Firth of Thames.

Brown Quail

Conspicuous colour Brown.
Habitat Inhabits rough country, which has plenty of vegetation cover such as manuka, broom, fern and gorse.
Range Found in the North Island from North Auckland to the Coromandel Peninsula, Bay of Plenty, King Country, Poverty Bay, Gisborne and Hawke's Bay. Absent from the southern parts of the North Island and the South Island.
Size 180 mm (House Sparrow 145 mm, California Quail 250 mm).
Status Introduced.

Family Phasianidae **Species** *Synoicus ypsilophorus*

Common names Brown Quail, Rat Quail

Description Crown Dark brown with a fine black spotting. Upperparts Brown mottled with black and chestnut. Throat Buff. Underparts Greyish-buff with black horizontal barring. Bill Bluish-grey. Legs and feet Yellow.

Conspicuous features Overall brown colouring is distinctive.

Conspicuous characteristics Has a small rat-like appearance, hence its nickname rat quail. When disturbed it scuttles away into roadside verges with head down. Often seen in family groups over the summer months. Flies with a fast whirr of wings. Lacks the topknot feathers of the California Quail.

Call Long drawn out, rising 'ker-wee ker-wee' whistle.

Nest On ground hidden in grass. Seven white, brown-speckled eggs are laid.

Target localities Bay of Islands district. Tiritiri Matangi Island.

New Zealand Pipit

Conspicuous colour Brown.
Habitat A bird of the open countryside but preferring wilderness, unkempt areas, to improved pasture. Prefers areas where scrubby roadside verges exist and roads are gravelled. Commonly found on beaches.
Range Found throughout New Zealand but seldom in the improved farming districts. Absent from forested areas.
Size 190 mm (House Sparrow 145 mm, Skylark 180 mm).
Abundance Uncommon.
Status Native.

Family Motacillidae **Species** *Anthus novaeseelandiae*

Common names Pipit, Pihoihoi

Subspecies Four are recognised: New Zealand Pipit (*A. n. novaeseelandiae*), Chatham Island Pipit (*A. n. chathamensis*), Auckland Island Pipit (*A. n. aucklandicus*), Antipodes Island Pipit (*A. n. steindachneri*).

Description Head Brown streaked with light brown and with a white eyebrow. Upperparts Brown streaked with light brown. Chin White. Breast White streaked with light brown. Underparts Whitish. Tail Dark brown with white outer feathers.

Conspicuous features The whitish eyebrow is noticeable, especially on birds in breeding plumage. White outer tail feathers are noticeable.

New Zealand Pipit

Conspicuous characteristics Has a tendency to bob the tail, especially upon alighting ✎ Has a tendency to signal its arrival by a loud, high-pitched 'scree' note ✎ Has a tendency to rise up in front of approaching traffic on dusty country roads and then fly in undulating swoops alongside or in front of cars.

Call A high-pitched 'tueet tueet' or 'scree scree' sound.

Nest On the ground hidden in grass. Four cream, brown-blotched eggs are laid.

Target localities Road to Cape Reinga ✎ Eglinton Valley near Te Anau.

Differences between the Pipit and the Skylark

- The Pipit lacks the Skylark's crest and rather square head.
- The Pipit is usually of upright stance, standing higher than the Skylark, which instead is inclined to sit back on its haunches with a sloping back posture.
- The Pipit has a habit of landing nearby and being friendly and confiding. At such times it will bob its tail vigorously.
- Pipits sing from a perch, although they will utter their 'scree' note when on the wing; Skylarks are vigorous aerial singers.

Starling (immature)

Conspicuous colour Brown.
Habitat A bird of towns, cities and open pasture.
Range Found throughout New Zealand away from forests and alpine areas.
Size 210 mm (House Sparrow 145 mm).
Abundance Common.
Status Introduced.

Family Sturnidae **Species** *Sturnus vulgaris* **Common name** Starling

Description – immature Upperparts Brownish-grey with some darker wing striping. Underparts Greyish. Eye Black with a dark-brown area in front of it and behind. Bill Brown.

Conspicuous features Dark eye on immature birds ✎ Brown bill.

Conspicuous characteristics Often seen with darker adult-plumaged birds in mixed flocks during January.

Call Raucous and noisy after leaving the nest.

Target locality Open countryside.

Song Thrush

Conspicuous colour Brown.
Habitat Vegetated areas from suburban gardens to hedged rural areas, exotic plantations and into native forest.
Range Found throughout New Zealand and the offshore islands.
Size 230 mm (House Sparrow 145 mm, Blackbird 250 mm).
Abundance Common.
Status Introduced.

Family
Muscicapidae

Species
Turdus philomelos

Common names
Song Thrush, Thrush

Description
Upperparts Olive-brown.
Underparts Brown speckled with white.
Bill Brown with a speck of yellow at the gape.

Conspicuous feature
Look for the brown breast streaking.

Conspicuous characteristics Feeds mainly on the ground where it takes worms, snails and insects Hops when feeding rather than walks Often heard breaking snails on hard surfaces Sometimes the female Blackbird, which is of brownish colouring, is mistaken for the Song Thrush.

Call Male birds start singing in the north of New Zealand at the end of April and continue through until January. Their song is often confused with that of the Blackbird. However, whereas the Blackbird sings in short phrases of a variety of notes, the Song Thrush repeats the same note, forming brackets, which it runs into the next bracket without a pause.

Nest Large bulky cup made of twigs and grasses bound with mud and lined with mud about 3 m up. Four greenish, black-spotted eggs are laid.

Target localities From gardens to forests.

Little Owl

Conspicuous colour Brown.
Habitat Frequents open farmland and arable land which has sufficient scattered vegetation or hedging to provide cover. Prefers hedging varieties such as pine (*Pinus radiata*) or macrocarpa (*Cupressus macrocarpa*).
Range South Island only with largest populations in Southland but fairly widely spread from Southland north to the Marlborough Sounds.
Size 230 mm (House Sparrow 145 mm, Morepork 290 mm).
Abundance Common in selected areas.
Status Introduced.

Family Strigidae **Species** *Athene noctua*

Common names Little Owl, German Owl

Description Head Grey-brown finely streaked with cream above eyes and on cheeks. Upperparts Brown-grey speckled with white. Underparts Brown-grey flecked with white vertical markings. Eyes Yellow.

Conspicuous features Smaller in size than the native Morepork • Yellow eyes.

Conspicuous characteristics When disturbed will fly out from cover and sit on a wire or a branch • Has an undulating flight pattern.

Call A high-pitched 'kiew-kiew'.

Nest In holes from the ground up, often in old barns. Four white eggs are laid.

Target localities Hagley Park, Christchurch • Fortrose district east of Invercargill • Winton district, north of Invercargill.

Blackbird (female)

Conspicuous colour Brown.
Habitat Gardens, parks and forest.
Range Throughout New Zealand.
Size 250 mm (House Sparrow 145 mm, Song Thrush 230 mm).
Abundance Common.
Status Introduced.

Family Muscicapidae **Species** *Turdus merula* **Common name** Blackbird

Description – female Upperparts Dark brown. Chin Grey. Breast and underparts Light brown speckled with dark brown. Bill Orange in breeding season, otherwise brown.

Description – immature Upper and underparts Dark brown with some breast speckling.

Description – male Upperparts and underparts Black. Bill Bright yellow in the breeding season. Eye-ring Yellow.

Conspicuous features Yellow eye-ring and bill.

Conspicuous characteristics Commonly feeds on lawns and pasture. When on the ground it hops rather than walks. Birds moult heavily over late summer, more so than the Song Thrush.

Call A 'tok tok tok tok' alarm call can be heard at any time of year.

Nest A bulky cup of grass, twigs and leaves bound with some mud and lined with fine grasses. Up to four greenish, brown-speckled eggs are laid. Both birds help with nest building and the feeding of young.

Target localities Gardens and lawns. Found almost everywhere but not above the snow line.

California Quail (female)

Conspicuous colour Brown.
Habitat A bird of rough scrubland, and pastureland that contains wilderness areas.
Range Found in both islands away from dense forest and away from closely settled pastoral land. Absent from Fiordland and some parts of Westland.
Size 250 mm (House Sparrow 145 mm).
Abundance Common in selected areas.
Status Introduced.

Family Phasianidae

Species
Callipepla californica

Common name
California Quail

Description – female
Head, breast and upperparts Tending to brown. Crest feather Shorter than male and brown. Underparts Brown.

Description – male
Forehead Grey to the eye. Crown Black, with a white edge band and a black crest feather which leans forward. Face and chin Black, edged with a white-curved throat collar. Nape Black and white spotted. Wings Dark grey. Breast Blue-grey. Underparts Buff, feathers edged with black to give a horizontal scaly effect. Chestnut wash to middle abdomen. Bill Black. Legs and feet Grey.

Conspicuous features Topknot feather / Scaly pattern on abdomen.

Conspicuous characteristics Usually to be found in pairs or coveys / When disturbed birds run and then fly off with whirr of wings.

Call A quite 'kik kik' seldom heard.

Nest A hollow in grasses on the ground. Up to 13 cream, brown-blotched eggs are laid.

Target localities Areas of scrubland / Roadside verges with thick scrub cover / Wenderholm Regional Park, Waiwera / Aroha Island, Kerikeri.

Spotted Dove

Conspicuous colour Brown.
Habitat A bird of old gardens, parks, and forest verges.
Range Common in many suburbs of Auckland and now slowly spreading both north to Warkworth and south to Tauranga.
Size 300 mm (House Sparrow 145 mm).
Abundance Common in selected areas.
Status Introduced.

Family Columbidae **Species** *Streptopelia chinensis*

Common names Spotted Dove, Lace-necked Dove, Malay Spotted Dove

Description Head Grey. Nape Black spotted with delicate white spots. Upperparts Brown. Underparts Brownish-grey. Tail Brown with outer feathers white.

Conspicuous features Undertail black and white in flight ✎ Spotted neck-collar.

Conspicuous characteristics A fast-flying dove which shows a long tail when in flight ✎ Feeds especially on gravel driveways and on mown lawns ✎ Calls from rooftops or chimneys ✎ Commonly seen at dusk carrying out display flights, swooping steeply upwards and then stalling, dropping downwards before swooping up once again.

Call A tri-syllabic 'ku kuu kuk' distinguishes it from the Barbary Dove which is bi-syllabic. Both species can be mistaken for the evening calls of Moreporks, which are also heard in some Auckland suburbs.

Nest A twiggy structure at about 4 m. Two white eggs are laid.

Target localities Cornwall Park, Greenlane, Auckland ✎ Puketutu Island across the causeway from the Mangere sewage plant ✎ Point View Drive, Howick ✎ Clevedon, east of Papakura.

Australasian Harrier

Conspicuous colour Brown.
Habitat A bird of the open countryside, straying over towns and cities.
Range Found throughout New Zealand and on offshore islands.
Size 550 mm (House Sparrow 145 mm, New Zealand Falcon 430 mm).
Abundance Common.
Status Native.

Family Accipitridae **Species** *Circus approximans*

Common names Harrier, Hawk, Kahu

Description Upperparts Dark brown. Underparts Reddish-brown streaked with dark

brown. Rump White.
Eye Yellow with brown centre and very yellow in male birds. Legs and feet Yellow.

Conspicuous features Light colouring of underparts ✎ Whitish rump on birds in the air ✎ Older birds are paler.

Conspicuous characteristics Flies low and level at first light ✎ Soars, glides and rises on thermals during daytime ✎ During the mating season makes 'scree'-type calls ✎ Regularly disturbed on roadsides, eating road-killed rabbits or possums ✎ Young birds often follow parents in level flight, calling plaintively.

Call The mating 'scree' calls are common over the spring months.

Nest Platform of grass or rushes, usually located in a swamp. Up to five white eggs are laid.

Target localities Open countryside throughout the country.

Differences between the Harrier and New Zealand Falcon

- The Harrier is the common raptor of the open New Zealand countryside; the New Zealand Falcon seldom ventures out over farmland in the North Island but does in eastern parts of the South Island high country.
- The Harrier is much larger than the New Zealand Falcon – 550 mm compared with 430 mm.
- The Harrier constantly soars and rises on thermals; the New Zealand Falcon is usually seen flying in a direct manner with a flight pattern similar to a New Zealand Pigeon, although it will briefly rise on thermals.
- The Harrier has a yellow eye; the New Zealand Falcon has a brown eye.

Pheasant (female)

Conspicuous colour Brown.
Habitat Prefers a wooded environment with plenty of cover.
Range Ranges throughout the North Island away from forests, and in the Canterbury areas of the South Island.
Size 800 mm (House Sparrow 145 mm).
Abundance Common.
Status Introduced.

Family Phasianidae **Species** *Phasianus colchicus* **Common name** Pheasant

Description – female Upper and underparts Buff-coloured.

Description – male Face Red. Head Black with bluish tinge on neck. Neck-collar White. Tufts Two short, black ear-tufts are on each side of the crown. Wings Grey-brown. Back Golden-brown with black speckling. Rump Golden-brown. Tail Buff, striped horizontally with up to 16 separate black bars. Underparts Golden-buff with some black speckling. Bill Light brown. Legs Grey.

Conspicuous features Dull-brown colouring.

Conspicuous characteristics Often seen gliding across gullies When disturbed in grass it rises quickly, with very fast wing-beats.

Call Usually silent.

Nest Bowl in grass on the ground. Up to nine olive-brown eggs are laid.

Target localities Open countryside with hedges and trees.

Fernbird

Conspicuous colour Brown.
Habitat Wet, swampy areas and semi-wet tidal verges on the main islands. Also in low fern and rush, away from wet areas, such as on Great Barrier Island.
Range In the North Island found from Tongariro north and on Great Barrier Island. In the South Island found from northeast Marlborough down the west coast to Otago and Southland but not in Fiordland. Also found on Stewart Island.
Size 180 mm (House Sparrow 145 mm).
Abundance Common in selected areas but secretive.
Status Endemic.

Family Sylviidae

Species *Bowdleria punctata*

Common names
Fernbird, Matata

Subspecies
Five are recognised: North Island Fernbird (*B. p. vealeae*), South Island Fernbird (*B. p. punctata*), Stewart Island Fernbird (*B. p. stewartiana*), Codfish Island Fernbird (*B. p. wilsoni*), Snares Island Fernbird (*B. p. caudata*).

Description Upperparts Brown streaked with dark brown. Eyebrows White. Underparts White streaked with brown from throat to abdomen.

Conspicuous features Spiky fern-like tail / Vertical brown streaking on breast.

Conspicuous characteristics Often does short, low flights above the vegetation / Tail is held in a downwards position when in flight / Birds often just arrive and peer from the vegetation at head height / Feeding birds have a habit of fossicking quickly through the vegetation, appearing and disappearing at speed.

Call A feeble 'tick tick tick' or 'uu-tik uu-tik' or a melodic 'too-tok too-tok' with the last note on a descending scale. Pairs are known to duet with one bird singing the first note and the other the second.

Nest A cup of woven reeds and grasses among rushes. Four pale-pink, brown-blotched eggs are laid.

Target localities Tiritiri Matangi Island / Tokaanu district, Lake Taupo / Sinclair Wetlands, Otago.

Marsh Crake

Conspicuous colour Brown.
Habitat Swamps, wetlands and raupo margins of lakes, similar to the habitat of the Spotless Crake.
Range Throughout the South Island. In the North Island it is known from Northland through Waikato to Bay of Plenty, and from Hawke's Bay south. New localities are regularly being discovered.
Size 180 mm (House Sparrow 145 mm, Spotless Crake 200 mm).
Abundance In selected areas but secretive.
Status Native.

Family Rallidae **Species** *Porzana pusilla*

Common names Marsh Crake, Koitareke, Baillon's Crake

Description — mature Face, throat, breast and abdomen Blue-grey. Crown and upperparts Chestnut-brown. Wings and back Chestnut-brown, with black-and-white mottling on secondary feathers. Flanks Horizontally streaked with black and white. Bill, legs and feet Greenish.

Description — immature Upperparts Brown streaked with dark brown. Underparts Light brown horizontally streaked with white.

Conspicuous features Smaller than the Spotless Crake · Barring and white flecking on flanks and wings · Paler than the Spotless Crake.

Call A harsh 'krek krek krek' is often heard at night. Also a harsh trill. Will respond to Spotless Crake taped calls.

Nest Platform of grasses and rushes on ground in wet area. Six olive-brown eggs are laid.

Target localities Lake Ngaroto boardwalk, Waikato · Sinclair Wetlands, Otago · Bushy Point, Invercargill.

Australasian Little Grebe

Conspicuous colour Brown.
Habitat Clean-water lakes and sand-dune lakes.
Range Found in limited areas, in the North Island north of Auckland and in the South Island in the South Canterbury district.
Size 250 mm (House Sparrow 145 mm, Dabchick 280 mm).
Abundance Common in a few selected areas.
Status Native.

Family Podicipedidae **Species** *Tachybaptus novaehollandiae*

Common name Little Grebe

Description – mature Crown, hindneck and upperparts Dark olive-brown. Sides of face and neck Dark brown with a rusty sheen. Upper and underwings White with brown tips to primaries. Underparts Light brown. Eye and tear drop Pale yellow.

Description – immature Crown Dark brown. Upperparts Dark brown. Side of face, neck and underparts Light brown.

Conspicuous features Slightly smaller than a Dabchick ✎ Yellow teardrop of naked skin just in front of eye.

Conspicuous characteristic Constantly dives.

Target locality Lake Waiporohita near Karikari, Northland ✎ Lake Rotokawau, near Awanui, Northland ✎ Lake Kereta, end of Wilson Road off South Head Road, north of Helensville.

Dabchick

Conspicuous colour Brown.
Habitat Deep and clean lakes, although sometimes found on shallower farm ponds and sand-dune lakes.
Range North Island only.
Size 290 mm (House Sparrow 145 mm, Australasian Little Grebe 250 mm).
Abundance Common in selected areas.
Status Endemic.

Family Podicipedidae **Species** *Poliocephalus rufopectus*

Common names Dabchick, New Zealand Dabchick, Weweia

Description – mature

Head and neck Blackish, finely streaked with silver. Upperparts Blackish-brown. Upper and underwings White with brown tips to primaries. Neck and upper breast Dark chestnut. Lower breast Silvery with brownish flanks. Bill Black. Eye Pale-yellow iris with black pupil. Legs and toes Blackish, with flat, lobed toes.

Description – immature Upper and underparts Brown but paler than mature bird.

Description – chick Upperparts Horizontally zebra-striped in black and white. Underparts White.

Conspicuous features Eye button-like and pale yellow around a black pupil Dark-chestnut breast on mature birds Underwing white on birds in flight.

Conspicuous characteristics Often viewed alongside Scaup Sits lower in the water than Scaup and has longer and more slender neck Lacks the longer tail of the Scaup Constantly diving, often surfacing some distance away Brings up food and feeds young above the water.

Call Usually silent, except for communication calls during the breeding season.

Nest A pile of rushes floating on the water but anchored. Three white eggs are laid.

Target localities Straka Refuge, Weranui Road, Waiwera Lake Rotorua near the Government Gardens, Rotorua Matata Lagoon, Bay of Plenty.

Banded Rail

Conspicuous colour Brown.
Habitat A bird of wet gullies, swamps and lake verges. Coastal populations live in mangrove estuaries and among other types of estuary vegetation.
Range In the North Island in areas north of Aotea Harbour on the west coast and Ohiwa Harbour in the Bay of Plenty on the east. In the South Island, found in the north-west corner only. Found on Stewart Island.
Size 300 mm (House Sparrow 145 mm, Black Petrel 460 mm).
Abundance Uncommon.
Status Native.

Family Rallidae

Species *Rallus philippensis*

Common name Banded Rail, Buff-banded Rail (Australia), Moho-pereru

Description Crown Chestnut. Face and throat Prominent white stripe above eye. Chestnut stripe through eye with grey chin and throat. Upperparts Olive-brown, marked with black and white. Breast Horizontally striped with black and white with a buff band across the middle. Underparts Horizontally barred with black and white. Undertail Buff with horizontal black bars widely spaced. Bill Brown.

Conspicuous features White eye-stripe above the eye which extends a short way down the side of the neck ✎ Buff-coloured breast band ✎ Black and white barring on the abdomen.

Conspicuous characteristics Feed actively with heads down ✎ Often chase each other when feeding ✎ Disturbed birds quickly run for cover with heads down ✎ When feeding birds often walk with a measured gait.

Call A high-pitched 'brururururu' sound often followed by 'tick tick tick' calls. A harsh 'craak' is sometimes heard which sounds like a fence wire being stretched through rusty post staples.

Nest A cup-shaped bowl in grasses or rushes, well hidden. Up to five pinkish, brown-blotched eggs are laid.

Note Races of this bird are also found through South East Asia, Australia, and the Pacific islands east to Tahiti.

Target localities Aroha Island near Kerikeri ✎ Wenderholm Park near Waiwera ✎ Miranda near the Pukorokoro River ✎ Opoutere Estuary, north of Whangamata ✎ Matata Lagoons, Bay of Plenty ✎ Waimea Inlet near Nelson ✎ Marahau Causeway after Kaiteriteri Beach.

New Zealand Scaup (female)

Conspicuous colour Brown.
Habitat Deep, clean, freshwater lakes throughout New Zealand including those bordering the coastal sand dunes.
Range Found through many parts of New Zealand where there are lakes. Common in the Rotorua and Taupo areas. Well spread on the lakes of the South Island.
Size 400 mm (House Sparrow 145 mm, Mallard 580 mm).
Abundance Common.
Status Endemic.

Family Anatidae **Species** *Aythya novaeseelandiae*

Common names Scaup, Black Teal, Papango

Description – female Head and neck Dull brown. Back Black. Upper and underwings White with dark primary tips. Underparts Dull brown. Eye Brown. Bill Blue-black, with a white crescent-marking at base of bill.

Description – male Upperparts and underparts Glossy black. Upper and underwings White with black primary tips. Eye Iris yellow. Bill Blue-black.

Conspicuous features White on wings in flight ✎ Yellow iris on male birds.

Conspicuous characteristics Flight is fast and just above the water ✎ When not swimming together in breeding pairs, birds are usually found in large loafing flocks in sheltered water ✎ When feeding, birds are continually diving and reappearing at about 15-second intervals ✎ Young birds dive along with their parents.

Call When sitting loafing in rafts, male birds continually utter a high-pitched, rattled whistle.

Nest A bowl of grass or rushes, lined with down, and close to the water. Up to eight cream eggs are laid.

Target localities Lake Rotorua, Bay of Plenty ✎ Lake Wakatipu at Queenstown.

Grey Teal

Conspicuous colours Brown and white.
Habitat Coast, estuaries, lakes and rivers.
Range Throughout New Zealand.
Size 430 mm (House Sparrow 145 mm, Mallard 580 mm).
Abundance Common.
Status Native.

Grey duck (for comparison)

Family Anatidae

Species *Anas gracilis*

Common names Grey Teal, Tete

Description Head Top of head is dark brown. Face and under neck Pale yellow. Upperparts and underparts Overall grey-brown appearance with feathers having dark brown centres and buff edges. Upperwings Brown, with a large black speculum which has a metallic green sheen, and which is edged with white, the leading white edge being triangular in shape. Underwing Brownish-grey. Eye Iris red. Bill Blue-grey upper mandible and yellowish lower. Legs and feet Greyish-brown.

Conspicuous features Considerably smaller than the Mallard ✎ Sexes are alike ✎ Pale-yellow face lacks the heavy black eye-stripe of the Grey Duck ✎ Looks smaller, neater and softer in colouring than the Grey Duck ✎ The white on the upperwings is very visible on in-flight birds.

Conspicuous characteristics Roosting birds sometime show white flank markings ✎ When flying in flocks they have an immaculate and coordinated flight pattern with individual wing-beats appearing to be in unison ✎ The wing-beats of Grey Teal are faster than either Mallard or Grey Duck.

Call Male birds have a loud sharp whistle, female birds a 'cuck cuck cuck' quacking sound.

Nest Bowl of grass lined with down in hole of tree or among tall grasses. Up to seven cream eggs are laid.

Target localities Miranda, Firth of Thames ✎ Many countryside lakes and wet areas.

Brown Teal

Conspicuous colour Brown.
Habitat Remnant populations are generally confined to muddy tidal-river estuaries, ponds and wet areas.
Range Great Barrier Island contains most birds. Isolated groups live in areas of Northland, on Tiritiri Matangi Island and on Little Barrier Island.
Size 480 mm (House Sparrow 145 mm, Grey Teal 430 mm).
Abundance Rare.
Status Endemic.

Family Anatidae **Species** *Anas aucklandica*

Common names Brown Teal, Pateke

Description – male Head Dark brown with a cream eye-ring around a black eye. Neck Brown with an indistinct white collar on front of bird. Upperparts Dark brown to brown. Wings Brown, with buff edges to feathers. Speculum Black with greenish sheen and white edges. Breast Chestnut. Underparts Buff. Flank spot White. Bill Blue-black.

Description – female Upperparts Dark brown with buff edges to feathers. Underparts Dull brown.

Conspicuous features Slightly larger than the Grey Teal Cream eye-ring Blackish speculum edged with white on in-flight birds White flank marking on male birds.

Conspicuous characteristics Birds loaf in large groups by day Birds become active towards dusk Mainland birds tend to be nocturnal feeders. Great Barrier Island birds feed throughout the day but are more active at dusk.

Call The male bird has a high-pitched, two-syllable 'wheeze'. The female has a dull, rapidly repeated quack.

Nest In a rush bush or tall grass on the ground. Four cream eggs are laid.

Target localities Teal Bay, near Helena Bay, north of Whangarei Tryphena, Great Barrier Island Tiritiri Matangi Island.

New Zealand Shoveler (female)

Conspicuous colour Brown.
Habitat Ponds and lakes, especially those which have indented rush and raupo-covered shorelines. They are also found on slow-moving rivers, on estuaries and muddy coastal strips.
Range Evenly spread through the North and South Islands. Reaches some of the higher altitude lakes, especially where seclusion is guaranteed.
Size 490 mm (House Sparrow 145 mm, Mallard 580 mm).
Abundance Common.
Status Endemic.

Family Anatidae

Species *Anas rhynchotis*

Common names Shoveler, Spoonbill, Kuruwhengi

Description – female in breeding plumage Upperparts and underparts Brownish. Upperwing Bluish-grey panel, white bar, green speculum and dark-grey primaries. Bill Grey.

Description – male in breeding plumage Head and neck Bluish-grey, with black behind bill and a white crescent-shaped stripe in front of a golden eye. Back Brown. Wings Bluish-grey panel, white bar, green speculum and dark-grey primaries. Breast Mottled brown and white. Abdomen Chestnut-brown with a white flank-spot. Bill Grey. Legs and feet Orange.

Description – eclipse Upperparts and underparts Drab brown, although male birds retain some blue-grey tonings to the head. White flank-spot is absent.

Conspicuous features Spoon-shaped bill ✎ In breeding plumage the male bird is brightly coloured ✎ White flank-spot.

Conspicuous characteristics Birds sit lower in the water than Grey Ducks or Mallards ✎ Appearance is of a short neck and big head ✎ Birds swim in a hunched position ✎ Tend to stay together when in a group of waterfowl ✎ Regularly swim in pairs ✎ Fast-flying, similar to Grey Teal; In-flight birds show bright wing-markings ✎ Birds prefer still water where water insects and larvae can safely breed ✎ Seldom seen away from water, unlike Grey Ducks and Mallards. This is because the heavy pasture seeds and shoots which Grey Ducks and Mallards eat are not able to be handled by the fine-edged, sieve-like spoon bills of the Shoveler.

Call Female a 'cuck cuck cuck', similar to the Grey Teal. Male bird, a 'clonk' sound.

Nest A bowl lined with down in thick grass or rushes. Four cream eggs are laid.

Target localities Straka Refuge, Weranui Road, Waiwera ✎ Waimeha Lagoon, Waikanae.

Nankeen Night Heron

Conspicuous colour Brown.
Habitat River and lake verges.
Range In New Zealand a small population breeds near Pipiriki on the Whanganui River. This species is widely spread from Indonesia to Australia and New Caledonia where it inhabits swamp areas, lake verges and river boundaries.
Size 560 mm (House Sparrow 145 mm, White-faced Heron 670 mm).
Abundance Uncommon.
Status Native.

Family Ardeidae **Species** *Nycticorax caledonicus*

Common name Nankeen Night Heron

Description – mature Crown and nape Blue-black. Two white plumes protrude from the nape. Face Greenish in front of eye and a white line above and behind eye. Back, neck and wings Chestnut-brown. Throat and neck White, with a pale-chestnut wash on upper neck. Underparts White. Iris Yellow. Bill Black, of short and stout appearance. Legs Yellow.

Description – immature birds Under and upperparts Brown, streaked with white.

Conspicuous features Brown colouring ✎ Yellow eye and legs.

Conspicuous characteristics Has a hunched posture ✎ Has a squat appearance when feeding ✎ Regularly stands before walking slowly ✎ Roosts by day; ventures out and starts feeding before dark, especially when feeding young.

Call Harsh croaks or quacks are often heard when the bird is feeding. The young when in the nest have a noisy cackle.

Nest A platform of loose sticks in a tree usually high up. Up to four pale greenish-blue eggs are laid.

Target locality Pipiriki on Whanganui River.

Grey Duck

Conspicuous colour Brown.
Habitat Generally found in secluded waterways in remote areas on lakes and rivers but some can be found in the headwaters of estuaries. Seldom seen in lowland wetlands where Mallards are common, possibly because Mallards take a wider range of food types.
Range Found throughout the North, South, and Stewart Islands. Also on the Chatham Islands and many offshore islands.
Size 580 mm (House Sparrow 145 mm, Mallard 580 mm).
Abundance Common.
Status Native.

Family Anatidae
Species *Anas superciliosa*
Common names
Grey Duck, Parera, Black Duck (Australia)
Description Crown Black. Face and throat Pale yellow with black eye-stripe running through the eye from bill to back of head. A similar but lesser dark stripe crosses the lower cheek. Upperparts and underparts Brown, feathers edged with a pale buff. Wings Brown with green speculum edged front and back with black. Bill Grey with dark tip. Legs and feet Greenish-brown.

Conspicuous features Yellow face with black eye-stripe ✎ Mallard size ✎ In flight looks black.

Conspicuous characteristics Usually found in pairs ✎ Tends to skulk in rushes at the edge of lakes and ponds ✎ Uncommon in parks and gardens, unlike the Mallard ✎ Rises slowly into the air when disturbed on a waterway ✎ Pairs in remote places are always Grey Ducks.

Note Grey Duck-Mallard hybrid birds are known in some areas.

Call Male bird has a high-pitched 'quek quek quek' call, and often a piping whistle, while the female makes a raucous 'quack quack'.

Nest On the ground, made of grass and sticks and lined with down. Up to 12 green eggs are laid.

Target localities Straka Refuge, Weranui Road, Waiwera ✎ Lake Whangape, Rangiriri, North Waikato ✎ Lake Gunn, Eglinton Valley, Te Anau ✎ All Day Bay Lagoon, Oamaru.

Mallard (female)

Conspicuous colour Brown.
Habitat Found in both fresh and saltwater environments.
Range Found throughout New Zealand and on offshore islands.
Size 580 mm (House Sparrow 145 mm, Grey Duck 550 mm).
Abundance Common.
Status Introduced.

Family Anatidae **Species** *Anas platyrhynchos* **Common name** Mallard

Description – female Head Brown with dark-brown eye-stripe. Upper and underparts Streaked and spotted brown and buff. Wings Brown with blue speculum. Bill Orange-brown.

Description – male Head Dark, glossy green. Breast Chestnut. Upper and underparts Grey. Wings Grey. Speculum Blue. Bill, legs and feet Orange.

Description – eclipse bird Upper and underparts Male birds attain feather patterns similar to female birds.

Conspicuous features Female Mallards, and hybrids between Mallards and Grey Ducks, lack the yellow and black face markings of Grey Ducks. Mallards appear lighter in eclipse plumage than Grey Ducks.

Conspicuous characteristics Found near human habitation such as on park lakes. Mallards usually take to the air quicker than Grey Ducks when disturbed. Head bobbing, neck stretching and circling of the male by the female on the water is noticed as birds pair-bond after February.

Call The male makes a high-pitched 'quek quek' call or just a piping whistle. The female makes a raucous quack.

Nest In grass or rushes, a bowl lined with moss. Up to 12 greenish eggs are laid.

Target localities Parks, public gardens and lakes.

Glossy Ibis

Conspicuous colour Brown.
Habitat Shallow freshwater areas such as the edges of lakes, rivers and swamps in inland situations.
Range Can appear in any locality that has swamps or lakes. Consistently turns up in North Waikato near Rangiriri and Whangamarino.
Size 600mm (House Sparrow 145 mm, White-faced Heron 670 mm).
Abundance Uncommon but a frequent visitor, most arriving in the month of November.
Status Native.

Family Threskiornithidae
Species *Plegadis falcinellus*
Common name Glossy Ibis
Description
Upper and underparts Brown with a greenish gloss on the wings.
Bill Olive-brown and markedly curved downwards.
Legs and feet Olive-brown.
Conspicuous features
Long, curved bill · About Cattle Egret size.
Conspicuous characteristics
In-flight shows long, curved bill · Flies with neck and legs extended · Tends to be very nervous · When feeding tends to disappear into the grass and rushes.

Target localities Rangiriri, Whangamarino in places where there are wetlands, lakes or water lying · Patetonga Lagoon, Hauraki Plains · All Day Bay Lagoon, south of Oamaru.

Australasian Bittern

Conspicuous colour Brown.
Habitat A bird of wetlands and wetland verges and sometimes open pasture.
Range Throughout New Zealand. More likely to be encountered in Northland, South Auckland and the Waikato.
Size 710 mm (House Sparrow 145 mm, White-faced Heron 670 mm).
Abundance Uncommon.
Status Native.

Family Ardeidae **Species** *Botaurus poiciloptilus*

Common names Australasian Bittern, Matuku-hurepo

Description Upperparts and underparts Brown, made up of both buff shadings and dark-brown shadings. Throat Whitish. Bill Upper mandible brown, lower mandible greenish. Legs Green.

Conspicuous feature Brown colouring.

Conspicuous characteristics When feeding, has a hunched posture and can spend many minutes motionless in this stance. When alerted, has a habit of freezing and extending bill skywards. When in this position the various shades of the brown feathering line up with surrounding reeds and rushes. Has a wafting flight of slow wing-beats, with head tucked in and legs trailing.

Call A strange vibrating booming. This sounds like a distant foghorn and is uttered throughout the breeding season. The female's reply is quieter. Note the booms are well spaced at 5–20-minute intervals.

Nest A bowl in rushes. Four olive-brown eggs are laid.

Target localities Lake Waahi, Huntly; Patetonga Lagoon, Hauraki Plains; Matata Lagoons, Bay of Plenty; St Anne's Lagoon, Cheviot, North Canterbury; Awarua Bay, Invercargill.

Black Swan (immature)

Conspicuous colour Brown.
Habitat Inland lakes, tidal harbours and estuaries.
Range Throughout New Zealand.
Size 1200 mm (House Sparrow 145 mm, Mute Swan 1500 mm).
Abundance Common.
Status Introduced.

Family Anatidae **Species** *Cygnus atratus* **Common name** Black Swan

Description – immature bird Upper and underparts Light brown. Upper and underwings White with brown on the secondary feathers only.

Description – mature Upper and underparts Black. Upper and underwings White with black on the secondary feathers only. Bill Red with a white band near the tip and a white tip. Legs and feet Black.

Conspicuous features A large black swan ✎ The 'painted' white band across the upper mandible is noticeable ✎ White wing primaries are noticeable.

Conspicuous characteristics When on the water birds continually raise their wings and flap them ✎ In flight, neck is outstretched ✎ Song and wing whistle are regularly heard overhead at night.

Call The Black Swan has a very musical song, usually heard when birds are flying. Also, a musical whistle is uttered when birds are sitting on the water.

Nest A mound of grass and rush stems near the shoreline and on the ground. Up to six pale-green eggs are laid.

Target locality Lake Pupuke, Takapuna, Auckland ✎ Lake Whangape, Rangiriri, North Waikato ✎ St Anne's Lagoon, Cheviot, North Canterbury.

Whimbrel

Conspicuous colour Brown.
Habitat Estuaries and harbours.
Range Can be expected in small numbers from the northern harbours of the North Island to Southland.
Size 430 mm (House Sparrow 145 mm, Bar-tailed Godwit 390 mm).
Abundance Rare.
Status Migrant.

Family Scolopacidae **Species** *Numenius phaeopus*

Common name Whimbrel

Subspecies Two are known to visit New Zealand: the Asiatic Whimbrel (*N. p. variegatus*) and the American Whimbrel or Hudsonian Curlew, (*N. p. hudsonicus*).

Description Head Brown with a white centre-stripe on crown and with white eyebrows. Upperparts Brown streaked with buff. Underparts Light brown streaked longitudinally with buff. Rump Whitish (Asiatic), light brown (Hudsonian). Tail Grey and black-barred. Bill Dark brown and curved downwards. Legs Bluish-grey.

Conspicuous features White head striping ✎ Down-curved bill.

Conspicuous characteristics Keeps to the edge of the godwit flocks when resting at high tide ✎ Remains alert when resting with head up.

Call A rippling trill of seven 'ti ti ti ti ti ti ti' notes. Birds will call throughout the night.

Target locality Miranda, Firth of Thames ✎ Matata Lagoons, Bay of Plenty ✎ Awarua Bay, Invercargill.

Eastern Curlew

Conspicuous colour Brown.
Habitat Found in estuaries and harbours.
Range Can be expected in any harbour or estuary.
Size 630 mm (House Sparrow 145 mm, Bar-tailed Godwit 390 mm).
Abundance Rare.
Status Migrant.

Family Scolopacidae **Species** *Numenius madagascariensis*

Common names Eastern Curlew, Long-billed Curlew

Description Upperparts Dark brown mottled with buff. Underparts Buff, longitudinally striped with dark brown. Eyebrow Whitish. Bill Dark brown, noticeably curved and darker on the tip. Legs Bluish-grey.

Conspicuous features Long down-curved bill ✎ Large size.

Conspicuous characteristics Generally remains apart from other birds on high-tide roosts ✎ Stays on the edges of flocks.

Call A long 'kroo-lee kroo-lee' uttered when a bird takes off.

Target localities Miranda, Firth of Thames ✎ Matata Lagoons, Bay of Plenty ✎ Awarua Bay, Invercargill.

Black-backed Gull (immature)

Conspicuous colour Brown.
Habitat Found around the coast and over pasture.
Range Found throughout New Zealand and offshore islands.
Size 600 mm (House Sparrow 145 mm, Pomarine Skua 480 mm).
Abundance Common.
Status Native.

See page 66 for adult birds.

Family Laridae

Species Larus dominicanus

Common name Black-backed Gull

Description — immature birds until year three Upper and underparts Dull brown streaked with dark brown. Bill and legs Brown.

Conspicuous features Overall brown.

Conspicuous characteristic The only big bird on most beaches.

Call A clear yodel of some volume.

Target locality Most beaches.

Brown Skua

Conspicuous colour Brown.
Habitat A coastal species during the breeding season. Ranges into deeper waters over the winter months.
Range The New Zealand subspecies (*Catharacta skua lonnbergi*) ranges around the south of the South Island, Stewart Island, Snares Islands and to the islands further south. It sometimes is seen around Cook Strait when not breeding.
Size 630 mm (House Sparrow 145 mm, Black-backed Gull 600 mm).
Abundance Common in selected areas.
Status Native.

Family Stercorariidae

Species *Catharacta skua lonnbergi*

Common names Brown Skua, Southern Skua, Subantarctic Skua, Hakoakoa

Description Upperparts and underparts Dark brown. Upper and underwings Dark brown with white wing-flashes on primary feathers.

Conspicuous feature White wing-flashes.

Call 'Charr-charr-charr' at nesting site only.

Nest Breeds either in loose colonies or as a solitary breeder, making a rough nest lined with grasses, twigs and dried seaweed placed on a promontory. Up to two light-brown, dark-brown-blotched eggs are laid.

Target localities Around Stewart Island.

Differences between a Brown Skua and an immature Black-backed Gull

- The Skua is heavier looking, has wider wings with very conspicuous white on them.
- The Skua has a thicker neck, and a much heavier bill with a hook on the end.

Northern Giant Petrel (mature)

Conspicuous colour Brown.
Habitat Sea.
New Zealand range Around New Zealand for most of the year.
Size 900 mm (House Sparrow 145 mm).
Abundance Common.
Status Native.

Family Procellariidae **Species** *Macronectes halli*

Common names Giant Petrel, Nelly, Pungurunguru

Description – mature Head Forehead pale brown with sides of face and chin grey. Upperparts Greyish-brown. Underparts Tending to dark grey. Underwings Brown with variable areas of grey. Bill Light tan with distinctive brown tip. Has a prominent nasal tube. Feet and legs Dark grey.

Description – immature Upper and underparts Black.

Conspicuous features Brownish colouring with lighter grey around the face, throat and upper breast ✎ Light-tan bill with brownish tip. This separates it from the Southern Giant Petrel which has a yellow bill tipped with green ✎ Heavy nasal tube.

Conspicuous characteristics Flight is often straight and direct ✎ Glides and wheels behind ships ✎ Can glide motionless for some distance.

Breeding islands close to New Zealand Chatham Islands, Stewart Island, Antipodes Islands, Campbell Island, Auckland Islands and Macquarie Island.

Breeding months August to February. Birds tend to nest in small, loose communities. They make a nest on a tussock bush and lay one white egg.

Range worldwide New Zealand west to South Africa.

Target localities Waters near Little Barrier Island ✎ Regularly seen from the Cook Strait ferry ✎ Regularly seen from both the Kaikoura coastline and from whale-watching excursions.

Kaka

Conspicuous colours Brown and red.
Habitat Forests.
Range Throughout New Zealand.
Size 450 mm (House Sparrow 145 mm, Red-crowned Parakeet 280 mm).
Abundance Uncommon.
Status Endemic.

Family Psittacidae **Species** *Nestor meridionalis* **Common name** Kaka

Subspecies Two are recognised: North Island Kaka (*N. m. septrionalis*), South Island Kaka (*N. m. meridionalis*).

Description Head Cap greyish on the North Island Kaka and grey on the South Island Kaka. Behind the eye is orange. Collar Dark crimson. Upperparts Olive-brown. Breast Olive-brown with reddish tonings. Abdomen and undertail Crimson. Underwing Red.

Conspicuous feature Large head of the in-flight bird.

Conspicuous characteristics Harsh calls of in-flight birds — Flying birds always appear to be poorly proportioned — Rounded wings — The sound of breaking sticks can indicate a feeding Kaka.

Call Harsh, guttural calls but also sweet musical notes.

Nest A hole in an old tree high up. Up to five white eggs are laid.

Target localities Tryphena, Great Barrier Island — Lake Waikareiti Track, Waikaremoana — Pureora Forest near Benneydale — Lake Gunn, Eglinton Valley — Stewart Island.

Fantail (breeding)

Conspicuous colours Brown and white.
Habitat Found anywhere vegetation is present, including gardens, parks and forests.
Range Throughout New Zealand and the offshore islands.
Size 160 mm (House Sparrow 145 mm, Grey Warbler 100 mm).
Abundance Common.
Status Native.

Family Monarchidae

Species *Rhipidura fuliginosa*

Common names Fantail (Maori names, which vary from one part of the country to the other, include Piwakawaka, Piwaiwaka, Tiwakawaka, Tiwaiwaka, Tirairaka.)

Phases Comes in both a pied and black form.

Subspecies Three are recognised: North Island Fantail (*R. f. placabilis*), South Island Fantail (*R. f. fuliginosa*), Chatham Island Fantail (*R. f. penita*).

Description – pied phase (breeding) Head Brownish-black with a white eyebrow. Upperparts Brown. Chin White, below which is a black bar. Underparts Tan colouring but yellowish when in full breeding plumage. Tail Two central feathers black, outer feathers white.

Description – pied phase (non-breeding) Upperparts Dull brown. Underparts Fawn. Chin Faint white line. Tail As for breeding bird.

Conspicuous features Breast takes on yellowish tonings in full breeding plumage ✎ During late summer, plumage a drab brown ✎ Fan-shaped tail, white and black.

Conspicuous characteristics Will enter houses and sit on lampshades ✎ Always active, constantly shifting position with sideways movement of tail.

Call A high-pitched 'cheet cheet' communication call. When in nesting territories male birds have a vocal and constant repetitive chattering call.

Nest Neat wine glass-shaped cup of twigs and leaves lined with moss. Four white, brown-speckled eggs are laid.

Target localities Found everywhere from gardens to forest.

Nankeen Kestrel (female)

Conspicuous colours Brown and white.
Habitat Pasture and wooded areas.
Range Selected areas.
Abundance Rare but a regular visitor.
Size 330 mm (House Sparrow 145 mm, New Zealand Falcon 430 mm).
Status Native.

Family Falconidae

Species *Falco cenchroides*

Common names
Nankeen Kestrel, Australian Kestrel

Description – female
Head Rufous with some black streaking. Upperparts Cinnamon brown, feathers spotted or streaked black. Underparts Light-brown breast to white abdomen. Tail Brown, barred with ten rows of black and a black tip to tail.

Description – male
Head and neck Blue-grey. Back and upperwings Cinnamon-brown with black primaries. Underparts White with a wash of buff on upper breast. Tail Grey with a black tip. Ceres (above bill) Yellow. Eye Brown with a thin, yellow eye-ring. Bill Horn-coloured with a black tip. Legs Yellow with black claws.

Conspicuous features Small size • Blue-grey head and tail on male bird.

Conspicuous characteristics Hovers in flight above prey before gliding down • Tends to fly with fast wing-beats and then soars with flat wings.

Call Shrill, excited chatter.

Nest In rock cavities, on cliff ledges or in old trees. Up to five buff, reddish-blotched eggs are laid. Nesting has not been recorded in New Zealand.

Target localities Could occur anywhere. Has been known from Naike, North Waikato and Te Mata Peak, Havelock North, Hawke's Bay.

Long-tailed Cuckoo

Conspicuous colours Brown and white.
Habitat Forests in summer but in parks and gardens in late autumn.
Range Throughout New Zealand in summer only.
Size 400 mm (House Sparrow 145 mm, Shining Cuckoo 160 mm).
Abundance Common.
Status A migratory native.

Family Cuculidae

Species *Eudynamis taitensis*

Common names
Long-tailed Cuckoo, Koekoea

Description
Upperparts Rich brown, barred with black.
Underparts Creamy-white, boldly streaked with brown and black longitudinal markings.
Tail Prominent horizontal barring on undertail and uppertail.

Conspicuous features One of the bigger birds of the New Zealand forest ✎ The long tail when in flight.

Conspicuous characteristics Screeching song uttered from a high perch ✎ Will spiral when in flight, often screeching at the same time ✎ Could be mistaken for a New Zealand Falcon when in flight ✎ Whiteheads, Yellowheads and parakeets will mob cuckoos if they enter their territories ✎ In late autumn can be found in suburbs and parks as it migrates north.

Call A long, harsh, drawn-out 'tuuueet', not unlike the introduced Greenfinch but on an ascending scale rather than a descending one, is the usual call. Also a musical song of ringing warbled, short notes.

Nesting Lays a creamy-white egg usually in the nest of the Whitehead, Yellowhead or Brown Creeper. Sometimes it also lays in nests of Silvereyes and Fantails.

Migration arrival and departure dates Arrives in New Zealand in early October and departs from February onwards.

Target localities Forest around Rotorua ✎ Pureora Forest, Benneydale, King Country ✎ Waikareiti Track, Lake Waikaremoana ✎ Pelorus Bridge, Marlborough ✎ Lake Manapouri, Manapouri ✎ Eglinton Valley, north of Te Anau.

Kookaburra

Conspicuous colours Brown and white.
Habitat Open country.
Range North Auckland. Was introduced from Australia in the 1860s but only a small number persist, mainly in the Waiwera to Kaukapakapa districts just north of Orewa.
Size 450 mm (House Sparrow 145 mm, Kingfisher 240 mm).
Abundance Uncommon.
Status Introduced.

Family Alcedinidae

Species *Dacelo novaeguineae*

Common name Kookaburra

Description

Head and underparts Whitish.
Wings Dark brown with some blue-grey flecking.
Tail Brown with darker brown barring.
Bill Large and black above and yellow below.

Conspicuous features

Larger than the Kingfisher in size ✎ Similar to the Kingfisher in shape and habit.

Conspicuous characteristics

Commonly sits on power wires or exposed branches of tall trees ✎ Regularly feeds on the ground.

Call Call is a raucous laugh – 'kuk kuk kuk and a he and a ho and a haw haw haw'. Heard mostly in the early morning and at sunset, but also through the day.

Nest A hole or cavity in an old tree. Up to three white eggs.

Target localities Glenbervie district, east of Whangarei ✎ Kawau Island out from Sandspit ✎ Wenderholm Regional Park, Waiwera, North Auckland ✎ Silverdale, North Auckland.

Peafowl (Peahen)

Conspicuous colours Brown and white.
Habitat Parks, gardens and open country.
Range Selected areas in Northland, South Auckland, Bay of Plenty, Poverty Bay, Hawke's Bay, Taranaki, northwest South Island and North Canterbury.
Size 900 mm (House Sparrow 145 mm, Arctic Skua 430 mm).
Abundance Uncommon.
Status Introduced.

Family Phasianidae

Species *Pavo cristatus*

Common name Peafowl

Description – female

Crown and neck
Metallic green. A tuft of fan-like feathers, tipped with green, protrudes from the back of head.

Face, chin and throat
White with brown eye-stripe.

Upperparts Brown.

Breast Buff-coloured.

Underparts Whitish.

Conspicuous feature
Head crest.

Conspicuous characteristics Very noisy at dusk as they gather together in trees for the night ✎ Of a gregarious nature ✎ Lost birds become agitated, calling madly.

Call Day call is a loud 'may-wee' screech. Evening call is more of a wail.

Nest A bowl in grasses or low vegetation. Up to six cream eggs are laid.

Target localities Waitangi Forest near Paihia in Northland ✎ Shakespear Regional Park at the end of Whangaparaoa Peninsula, North Auckland.

Australasian Crested Grebe

Conspicuous colours Brown and white.
Habitat Lakes.
Range South Island lakes.
Size 500 mm (House Sparrow 145 mm, Arctic Skua 430 mm).
Abundance Uncommon.
Status Native.

Family Podicipedidae **Species** Podiceps cristatus

Common name Crested Grebe (sometimes Great Crested or Southern Crested Grebe)

Description — breeding Crown Brown with tufts projecting upwards on the back of head. Face White. Neck A frill of rufous feathers hangs from under the cheeks. Foreneck White. Upperparts Brown. Underparts White. Bill Brown and finely pointed. Feet Greenish with flat-lobed toes.

Description — chicks Upperparts Zebra-striped in black and white horizontal markings. Underparts White.

Conspicuous features Sits low in the water and reveals a long thin neck ✎ Head is held at a right angle to the neck ✎ Long white neck ✎ White on forewing and primaries when in flight ✎ Retains adult plumage the year around.

Conspicuous characteristics Constantly diving and bobbing to the surface ✎ In flight long neck is held straight and slightly below body line ✎ Flies very close to the water ✎ Usually seen in pairs ✎ Young ride among folded wings of swimming parent ✎ Usually seen feeding near the rush line.

Call Usually silent but whistle type calls occasionally made in breeding season.

Nest A platform of floating sticks and reeds near the shoreline. Three white eggs.

Target localities Lake Forsyth, near Lake Ellesmere, Canterbury, in winter ✎ Lake Pearson and Grasmere, Arthurs Pass ✎ Lake Clearwater, Mt Somers district, Canterbury ✎ Lake Heron, Mt Somers district, Canterbury.

Red-necked Stint (non-breeding)

Conspicuous colours Brown and white.
Habitat Coast and estuaries.
Range Scattered throughout New Zealand.
Size 150 mm (House Sparrow 145 mm, Wrybill 200 mm).
Abundance Rare.
Status Migrant.

Family Scolopacidae **Species** *Calidris ruficollis*

Common name Red-necked Stint

Description – non-breeding Forehead and eyebrow White. Head and upperparts Grey with brownish tinge. Neck Traces of red on sides. Wings Greyish, with a white line across the middle of the wings. Tail Whitish outer feathers with black inner. Underparts White. Bill and legs Black.

Description – breeding Face, neck, chin and throat Brick red. New Zealand birds show only slight reddening prior to departure. Upperparts Dark brown with rufous edges to feathers. Underparts White.

Conspicuous features The smallest of the Arctic-breeding wading birds to come to New Zealand ✎ Bill is black, short and straight.

Conspicuous characteristics Prefers to feed on areas of mud covered by a thin film of water ✎ Feeds busily with sewing-machine action ✎ Runs here and there ✎ Usually found towards the edges of Wrybill flocks ✎ Often feeds in the company of other sandpipers and Wrybills ✎ In flight, birds fly low and direct in a compact group.

Target localities Miranda, Firth of Thames ✎ Maketu Harbour, Bay of Plenty ✎ Embankment Road, Lake Ellesmere, near Christchurch ✎ Waituna Lagoon near Invercargill.

Curlew Sandpiper (non-breeding)

Conspicuous colours Brown and white.
Habitat Coast and estuaries.
Range Scattered throughout New Zealand.
Size 190 mm (House Sparrow 145 mm, Wrybill 200 mm).
Abundance Rare.
Status Migrant.

Family Scolopacidae **Species** *Calidris ferruginea*

Common name Curlew Sandpiper

Description – non-breeding Face Greyish-brown with white eyebrow. Upperparts Grey-brown. Wings Brown, with a thin, white line across middle of wings. Rump White. Breast Greyish wash. Underparts and underwing White. Tail White with brown tip. Bill Black and noticeably down-curved at the tip. Legs Black.

Description – breeding (note that female birds are paler) Face White around base of bill and above eye, otherwise brick red. Head and neck Brick red. Back and wings Red, with flecks of black and silver. Breast and underparts Brick red with black barring. Rump, undertail and underwing White.

Conspicuous features Smaller than Sharp-tailed and Pectoral Sandpipers / Down-curved tip to the bill / Non-breeding birds have very white underparts / White rump, and white wing line on upperwings of bird in flight / Undergo a colour transformation into breeding plumage in January.

Conspicuous characteristics Usually feed among Sharp-tailed Sandpipers or Wrybill / Tend to start feeding immediately the tide has turned / Very fast and manoeuvrable fliers when in flocks.

Call A gentle 'chirrup' sound.

Target localities Miranda, Firth of Thames / Manawatu Estuary, Foxton.

Mongolian Dotterel (non-breeding)

Conspicuous colours Brown and white.
Habitat Coast and estuaries.
Range Consistently found at such places as Te Hihi on the Manukau Harbour, Miranda in the Firth of Thames, Ahuriri near Napier, Embankment Road and Kaitorete Spit at Lake Ellesmere, and Farewell Spit.
Size 200 mm (House Sparrow 145 mm, Wrybill 200 mm).
Abundance Rare.
Status Migrant.

Family Charadriidae **Species** *Charadrius mongolus*

Common names Mongolian Dotterel, Lesser Sand Plover

Description – non-breeding Forehead White extending to a white eyebrow. Upperparts Greyish-brown. Shoulder-tabs Greyish-brown. Underparts White. Eye Black with dark surrounding feathers. Bill Black. Legs Grey.

Description – breeding plumage Forehead White edged with black. Face and chin White with a black line through eye to behind. Upperparts Greyish-brown. Chin White faintly edged with black. Head, breast-band and nape Chestnut. Underparts White. Bill Black. Legs Grey.

Conspicuous feature Dark eye in breeding and non-breeding birds.

Conspicuous characteristics Runs, pauses and then dips its bill forward ✎ Associates with Banded Dotterel and Wrybill.

Target localities Miranda, Firth of Thames ✎ Ahuriri, Napier, Hawke's Bay ✎ Embankment Road and Kaitorete Spit, Lake Ellesmere ✎ Waituna Lagoon near Invercargill.

Differences between Mongolian Dotterel and Banded Dotterels – non-breeding

- Mongolian Dotterels have a brownish crown and head; Banded Dotterels have tawny tonings.
- Mongolian Dotterels have dark brownish eye surrounds; Banded are tawny around the eye.
- Mongolian Dotterels are darkish-brown on the upperparts; Banded tend to tawny colouring.

Mongolian Dotterel (non-breeding)

Branded Dotterel (non-breeding)

Differences between Mongolian Dotterels and Banded Dotterels – breeding

- Banded Dotterels lack the dark eye.
- Banded Dotterels have double bands - black upper and chestnut lower - on the breast with a clean, white barrier between. Mongolian Dotterels have a single wide chestnut coloured band.

Sharp-tailed Sandpiper

Conspicuous colours Brown and white.
Habitat Coast and estuaries.
Range Scattered throughout New Zealand.
Size 220 mm (House Sparrow 145 mm, Wrybill 200 mm).
Abundance Rare.
Status Migrant.

Family Scolopacidae **Species** *Calidris acuminata*

Common name Sharp-tailed Sandpiper

Description – non-breeding Crown Traces of rufous colouring over brown. Face Greyish with a white eyebrow. Upperparts Brown with buff edges to feathers. Wings Brown with a faint white line across middle of wing. Throat Brownish. Abdomen White. Tail Black outer, white inner, and brown centre feathers. Bill Brown slightly down-curved at tip. Legs Yellowish-green.

Description – breeding Crown Rufous. Head, neck and upperparts Rufous tonings over greyish-brown. Throat and upper breast Rufous. Abdomen White flecked with grey streaks.

Conspicuous feature Rufous crown.

Conspicuous characteristics Regularly seen on the edge of roosting Wrybill flocks. Tends to remain feeding while other species rest at high tides.

Target localities Miranda, Firth of Thames. Ahuriri, Napier, Hawke's Bay.

Differences between Sharp-tailed Sandpipers and Pectoral Sandpipers

- Sharp-tailed Sandpipers have a gradual blending of the brown of the breast into the white of the under parts.
- Pectoral Sandpipers have a sharp defined line between the two.
- Sharp-tailed Sandpipers become rusty coloured around throat and breast as they attain breeding plumage.
- Pectoral Sandpipers go a dark brown on throat and breast, and browner on the upper parts.

Pectoral Sandpiper

Conspicuous colours Brown and white.
Habitat Coast and estuaries.
Range Scattered throughout New Zealand.
Size 230 mm (House Sparrow 145 mm, Wrybill 200 mm).
Abundance Rare.
Status Migrant.

Family Scolopacidae **Species** *Calidris melanotos*

Common name Pectoral Sandpiper

Description – non-breeding
Head, throat and breast Brown streaked with dark brown. The brown of the breast ends abruptly in a clean line at the start of white. Upperparts Dark brown with buff edges to feathers.
Wings Dark brown and buff with a thin white line across the middle of wings. Abdomen White. Bill Brown with a yellow base. Legs Yellow.

Description – breeding Upperparts Dark brown with some chestnut and buff. Breast Brown heavily streaked with dark brown.

Conspicuous feature Has a sharp cut-off line between the brown of the breast and the white of the abdomen.

Conspicuous characteristics Usually to be found among flocks of Sharp-tailed Sandpipers ✎ Remains feeding at high-tide roosts while other species remain resting ✎ Often found on the edges of roosting Wrybill flocks ✎ Has a habit of stretching its neck forward while holding wings back.

Call Loud, harsh 'treet treet' sound.

Target localities Miranda, Firth of Thames ✎ Ahuriri, Napier, Hawke's Bay.

Terek Sandpiper

Conspicuous colours Brown and white.
Habitat Coast and estuaries.
Range Scattered throughout New Zealand.
Size 230 mm (House Sparrow 145 mm, Wrybill 200 mm).
Abundance Rare.
Status Migrant.

Family Scolopacidae **Species** *Tringa terek*

Common name Terek Sandpiper

Description – non-breeding Upperparts Brownish-grey. Eyebrows White. Breast Grey with darker shoulder-tabs. Abdomen White. Bill Brown with a yellow base and up-curved. Legs Yellow.

Conspicuous features Long up-curved bill ✎ Yellow legs.

Conspicuous characteristics Squat posture, standing on short legs ✎ Tends to bob both its head and its tail when it lands ✎ Feeds with agile running movements, sometimes very fast ✎ Will wade into shallow water when feeding ✎ Tends to associate with flocks of Wrybill roosting on the flock edge ✎ At high tide it usually roosts along the waterline ✎ When with Wrybills it looks a more solid bird.

Call A trilling 'turloop too'.

Target localities Miranda, Firth of Thames ✎ Ahuriri, Napier.

Large Sand Dotterel

Conspicuous colours Brown and white.
Habitat Coast and estuaries.
Range Scattered throughout New Zealand at selected harbours and estuaries.
Size 240 mm (House Sparrow 145 mm, Wrybill 200 mm).
Abundance Rare.
Status Migrant.

Breeding

Non-breeding

Family Charadriidae   **Species** *Charadrius leschenaultii*

Common names Large Sand Dotterel, Greater Sand Plover

Description – non-breeding Forehead White extending to above eye. Upperparts including shoulder-tabs Grey-brown. Underparts White. Bill and legs Black.

Description – breeding Crown, nape and breast Reddish-orange. Forehead White edged with black. Eye-stripe A black marking extends from bill through eye to back of head. Upperparts Brownish. Chin White. Underparts White. Eye Black.

Conspicuous features Dark eye ✎ Long legs.

Conspicuous characteristics Has a tendency to stand motionless ✎ Runs and pauses when feeding.

Call A short 'trrri' sound.

Target localities Jordans Road on Kaipara Harbour, north of Helensville ✎ Miranda, Firth of Thames.

Differences between Large Sand Dotterels and Mongolian Dotterels

- Large Sand Dotterels appear leggy and stand taller than Mongolians.
- Large Sand Dotterels have more horizontal posture than Mongolians.
- In breeding plumage, Large Sand Dotterels have no black on throat but have more orange on breast.

New Zealand Dotterel (non-breeding)

Conspicuous colours Brown and white.
Habitat Coast and estuaries.
Range North of the North Island and on Stewart Island.
Size 250 mm (House Sparrow 145 mm, Wrybill 200 mm).
Abundance Uncommon.
Status Endemic.

Family Charadriidae

Species *Charadrius obscurus*

Common names New Zealand Dotterel, Red-breasted Dotterel

Subspecies Two are recognised: Northern New Zealand Dotterel (Tuturiwhatu, *C. o. aquilonius*) and Southern New Zealand Dotterel (*C. o. obscurus*).

Description – non-breeding
Head Brown with a white forehead and a white eye-stripe. Upperparts Brown. Underparts White. Bill Black and robust. Legs Grey.

Description – breeding Breast Male bird attains a red breast and abdomen in July. Female bird attains lesser amounts of red colouring.

Conspicuous features A large bird compared with the more common Banded Dotterel ✎ White eye-stripe ✎ Dark eye and bill ✎ Red breast in breeding season.

Conspicuous characteristics Confiding and can be approached ✎ Runs and pauses when feeding ✎ A rounded and hunched bird when standing ✎ A strong flier which moves from high-tide roost to high-tide roost when not in a breeding territory ✎ Blends in well with the high-tide roost and breeding territory.

Call A 'kreek kreek' call when in flight.

Nest A scrape in the sand with minimal nest lining. Up to three buff-brown, dark-brown-blotched eggs are laid. In the North Island, nests are usually found behind the beach in locations with 360° views. Very often, mated pairs occupy territories at the opposite ends of beaches. On Stewart Island birds nest high on the island in prominent positions with good visibility, although there are records of sand-dune nests from Mason Bay.

Target localities Kawakawa Bay, east of Papakura ✎ Miranda, Firth of Thames ✎

Pacific Golden Plover (non-breeding)

Conspicuous colours Brown and white.
Habitat Coast and estuaries.
Range Scattered about New Zealand.
Size 250 mm (House Sparrow 145 mm, Wrybill 200 mm).
Abundance Common in selected areas.
Status Migrant.

Family Charadriidae **Species** *Pluvialis fulva*

Common names Golden Plover, Pacific Golden Plover

Description – non-breeding Head Brown with a white stripe above the eye. Throat Buff. Upperparts Brown with feathers edged with golden-buff. Breast Light brownish. Abdomen Brown but white undertail. Bill Black with white feathers around the base.

Description – breeding plumage Crown of head and upperparts Brown, heavily flecked with golden-yellow and white. Neck A white dividing line, starting at the forehead and travelling behind the eye and down the side of the neck, separates the black underparts from the golden upperparts. Underparts Black.

Conspicuous features Golden tonings ✎ Striking black underparts with white line from head to flanks on breeding birds.

Conspicuous characteristics Holds head high ✎ Runs and pauses when feeding

Will feed on pasture · Has an alert stance · At high tide will stand motionless for some time · Upon landing, birds hold their wings erect for a moment before folding.

Call Usually a clear two-syllabic 'tuill tuill' sound.

Target localities Aroha Island, Kerikeri · Jordans Road, Kaipara Harbour, north of Helensville · Clarks Beach and Te Hihi, Manukau Harbour · Waituna Lagoon, east of Invercargill.

Differences between Pacific Golden Plover (*P. fulva*) and American Golden Plover (*P. dominica*), which occasionally visits New Zealand and causes confusion

- Folded primary wing feathers of the American Golden extend well beyond the tail when the bird is standing; those of the Pacific Golden extend minutely past the tail.
- Folded tertial feathers of the American Golden lie well back from the tip of the tail; those of the Pacific Golden lie closer to the tip of the tail.
- American Golden Plovers are described as 'fairly bulky'; Pacific Golden Plovers are smaller and slimmer.
- American Golden Plovers in breeding plumage have more shoulder-white than the Pacific Golden.
- American in non-breeding plumage have a whiter eyebrow stripe than the Pacific with greyish head tonings.
- American have greyish-brown upperparts, spotted with pale yellow in non-breeding plumage; Pacific have brown upperparts spotted with bright yellow in non-breeding plumage.

Bar-tailed Godwit (non-breeding)

Conspicuous colours Brown and white.
Habitat Coast and estuaries.
Range Throughout New Zealand.
Size 390 mm (House Sparrow 145 mm, Wrybill 200 mm).
Abundance Common.
Status Migrant.

Family Scolopacidae

Species *Limosa lapponica*

Common names Bar-tailed Godwit, Eastern Bar-tailed Godwit, Kuaka

Description – non-breeding Upperparts Brown streaked with dark brown. Underparts White. Rump and tail White barred with brown. Bill Brownish-pink with a black tip and curved upwards. Legs Grey.

Description – breeding plumage (male) Head Brick-red with whitish stripe above the eye. Upperparts Dark brown with buff edges to feathers. Neck and underparts Chestnut. Underwing White. Tail Brown barred with white. Bill Brownish-pink.

Description – breeding plumage (female) Upperparts Greyish, streaked with brown. Breast Buff, lightly tinged with red. Underparts White with fine barring on edges of abdomen.

Conspicuous features A large wading bird ✎ Females are larger than male birds ✎ Bill is long and up-curved ✎ Females have longer bills than males.

Conspicuous characteristics Flies in loose and straggly skeins ✎ Congregates in large numbers on high-tide roosts ✎ Feeds busily once the tide recedes ✎ Digs bill deeply into mud to feed, so often seen with a muddy face ✎ Males develop rufous tonings towards the end of summer ✎ Auckland birds are often seen changing harbours in straggly skeins.

Call A soft 'kit kit kit' or a 'kew kew kew' sound. Just prior to migration birds become very excited and noisy at high-tide roosts.

Arrival dates From mid-September.

Departure dates Birds congregate in northern harbours from February onwards at places like Manukau Harbour and Miranda; start leaving mid-March.

Overwintering birds At any time of year, including winter, godwits in small numbers, often in breeding plumage, can be seen on many harbours.

Target localities Auckland's muddy beaches ✎ Kawakawa Bay, east of Papakura ✎ Miranda, Firth of Thames ✎ Motueka Estuary, Marlborough ✎ Fortrose Estuary, east of Invercargill.

Fluttering Shearwater

Conspicuous colours Brown and white.
Habitat Sea.
Range Around New Zealand.
Size 330 mm (House Sparrow 145 mm).
Abundance Common.
Status Endemic.

Family Procellariidae **Species** *Puffinus gavia*

Common names Fluttering Shearwater, Pakaha

Description Upperparts Dark brown. Underparts White. Bill Dark grey. Feet Brownish marked with white.

Conspicuous features A small shearwater ✎ White underparts ✎ Fine dark grey bill ✎ Feet protrude slightly behind the tail in flight.

Conspicuous characteristics Rapidly flutter their wings on take-off with feet assisting ✎ Through the months of mid-December, January and February come together in communal flocks and sit on the water in rafts ✎ Often follow schooling fish ✎ Flight pattern is four to five wing-beats and then a short glide ✎ Very similar to the Manx Shearwater (*P. puffinus*) of the northern hemisphere.

Call Very noisy near their breeding colonies, where they make a 'ka-haa ka- haa kehek' type call.

Breeding islands close to New Zealand Breeds from the Three Kings Islands in the north to islands in Cook Strait.

Breeding months September to February. One white egg is laid in a burrow.

Range worldwide Around New Zealand and to the southwest coast of Australia.

New Zealand range Particularly common in northern waters.

Target localities Waters from Sandspit ✎ Waters to Tiri Tiri Island ✎ Waters of the Hauraki Gulf.

Hutton's Shearwater

Conspicuous colours Brown and white.
Habitat Sea.
Range From Hawke's Bay south to Otago.
Size 360 mm (House Sparrow 145 mm, Fluttering Shearwater 330 mm).
Abundance Common in selected areas.
Status Endemic.

Family Procellariidae

Species *Puffinus huttoni*

Common names Hutton's Shearwater, Titi

Description Upperparts Dark brown. Underparts White. Bill Dark grey. Feet Brownish.

Conspicuous characteristics Often in rafts off the Kaikoura coast ✎ Mainland breeding high in the Kaikoura mountains ✎ In the breeding season will flock to about 1 km off the shoreline at dusk.

Mainland breeding site On the Seaward Kaikoura Range in the northern part of the South Island, above 1200 m.

Breeding months August to April. One white egg is laid in a burrow.

Range worldwide Around New Zealand and the southern half of Australia.

Call At breeding ground 'ko-oo ko-oo ko-oo ko – kee-kee-kee ah' is heard.

Target localities Waters from Cook Strait to Otago Peninsula ✎ Waters around Kaikoura.

Differences between Fluttering Shearwaters and Hutton's Shearwaters

- Fluttering Shearwaters are smaller in size, with shorter bills by 4 mm, have lighter feather colouration, and are whiter under the wings.
- Fluttering Shearwaters nest on small predator-free islands from the Three Kings to Marlborough Sounds in the south. Most burrows are at low altitudes. Hutton's Shearwater breeds on the mainland of the South Island high in the Kaikoura Ranges above 1200 m.
- Fluttering Shearwaters start breeding in mid-September (egg dates late September to mid-October); Hutton's start breeding mid-October (egg dates November).
- Hutton's Shearwaters are birds of the open seas rather than of waters around sheltered offshore islands and bays.

Arctic Skua (dark phase)

Conspicuous colours Brown and white.
Habitat Sea.
Range Around New Zealand.
Size 430 mm (House Sparrow 145 mm, Pomarine Skua 480 mm).
Abundance Uncommon.
Status Migrant.

Family Stercorariidae

Species *Stercorarius parasiticus*

Common names Arctic Skua, Arctic Jaeger

Phases The Arctic Skua comes in dark, light and variable phases. Because of this variation in colour the descriptions below are approximate.

Description – non-breeding (dark phase) Upperparts and underparts Brown, with crown darker. Wings Brown, usually with white-flashes at the base of the primary wing feathers on the underwings and with traces of white on the upperwings. The wing-flashes are sometimes missing.

Description – non-breeding (light phase) Crown, neck and upperparts Dark brown. Chin and throat White. Nape Yellow. Breast Brown band. Underparts White. Tail Brown, with a protruding central tail feather.

Conspicuous features Prominent tail feather but not on non-breeding plumage birds ✎ White underwing flashes.

Conspicuous characteristics Preys on fish caught by terns, although it also catches its own food ✎ Flies steadily and level when searching for feeding terns ✎ Extremely manoeuvrable when in flight chasing terns ✎ Sits on the water like a gull ✎ Has the habit of approaching fishing boats and sitting on the water nearby ✎ Will roost on beaches in the vicinity of other birds such as Caspian Terns.

Breeding localities Northern hemisphere, north of the 58th parallel, in loose colonies on offshore islands, open tundra and on open sandspits.

Migration Birds head to the southern hemisphere in July, south to the 46th parallel. They return north in March. Some immatures are known to stay on.

Target localities Waters out from Sandspit in late summer ✎ Miranda, Firth of Thames, near nesting terns.

Pomarine Skua (dark phase)

Conspicuous colours Brown and white.
Habitat Sea.
Range Around New Zealand.
Size 480 mm (House Sparrow 145 mm, Arctic Skua 430 mm).
Abundance Uncommon.
Status Migrant.

Family Stercorariidae **Species** *Stercorarius pomarinus*

Common name Pomarine Skua

Phases Dark and light.

Description – non-breeding (dark phase) Crown, neck and upperparts Dark brown. Nape Pale brown. Chin, throat and breast Brown. Underparts Brown. Tail Brown, with a slight protruding central tail feather more noticeable in breeding plumage birds.

Conspicuous features Prominent tail feather but not on juvenile birds ✐ Sometimes white flashes.

Conspicuous characteristics Preys on fish caught by terns, although it also catches its own food ✐ Flies steadily and level when searching for feeding terns ✐ Extremely manoeuvrable when in flight chasing terns ✐ Sits on the water like a gull.

Breeding localities Northern hemisphere north of the 58th parallel, in loose colonies on offshore islands, open tundra and on open sandspits.

Migration Birds head to the southern hemisphere in July, south to the 46th parallel. They return to the northern hemisphere in March. Some immatures are known to stay on.

Target locality Waters out from Sandspit in late summer.

Buller's Shearwater

Conspicuous colours Brown and white.
Habitat Sea.
Range Around New Zealand.
Size 460 mm (House Sparrow 145 mm, Fluttering Shearwater 330 mm).
Abundance Common.
Status Endemic.

Family Procellariidae

Species *Puffinus bulleri*

Common names Buller's Shearwater, Rako

Description Head Dark brown above the eye. Pure white under the eye. Upperparts Light grey. Upperwings Light grey with dark-brown, open M marking across wings and rump. Primaries Brown. Rump Dark brown. Underwings Clean white edged with black. Underparts Pure white. Tail Brown with a black tail tip. Bill Bluish-grey. Feet Bluish-grey.

Conspicuous features Slightly bigger in size than the Sooty Shearwater and the Flesh-footed Shearwater, both of which are common in the same waters off Northland Clean white underparts Brown, open M marking on the wings and rump.

Conspicuous characteristics Regularly comes in close to boats Often seen in rafts with Fluttering Shearwaters Will join up with feeding groups of Flesh-footed and Sooty Shearwaters Soars along waves close to water.

Major breeding islands near New Zealand Breeds only on the Poor Knights Islands off the east coast of Northland.

Breeding months October to May. Lays one white egg in a burrow.

Range worldwide To Australia and north in the Pacific to the North American coast.

Target localities Waters out from Sandspit Tiritiri Matangi Island passage.

Australasian Gannet (immature)

Conspicuous colours Brown and white.
Habitat Coastal areas and sea.
Range Around New Zealand.
Size 890 mm (House Sparrow 145 mm, Red-billed Gull 370 mm).
Abundance Common.
Status Native.

Family Sulidae **Species** *Morus serrator*

Common names Australasian Gannet, Takapu

Description – immature Upperparts Brown. Underparts Brown with varying amounts of white until years four when they resemble adults.

Description – mature Forehead White with black markings in front of eyes. Head Yellow. Upperparts and underparts White. Wings White with black-tipped primary feathers. Tail Black-tipped. Bill and bare skin of face Bluish-grey. Black line around gape. Feet and legs Greyish with feet striped with yellow.

Conspicuous features Brown on immature birds ✎ Yellow head on mature birds.

Conspicuous characteristics Birds feed by diving on to fish from considerable heights with wings folded back. If they catch a fish they usually bob to the surface and eat it before becoming airborne again. Otherwise they take to the air promptly ✎ Birds are regularly seen flying parallel to the coastline with heads down.

Call Excited, high-pitch chatter heard at the breeding colony.

Target localities Muriwai, west of Auckland ✎ Waters out from Auckland and Sandspit ✎ Cape Kidnappers, Hawkes Bay.

House Sparrow (male)

Conspicuous colours Brown, white and black.
Habitat Towns, suburbs and in rough open countryside.
Range Found throughout New Zealand and on some of the outlying islands.
Size 145 mm (Chaffinch 150 mm).
Abundance Common.
Status Introduced.

Family Ploceidae

Species *Passer domesticus*

Common name Sparrow

Description – male
Crown Grey.
Upperparts Brown.
Rump Greyish-brown.
Wings Brown, streaked black with a white wingbar.
Throat Black.
Underparts Greyish-white.

Description – female
Upperparts Brown, with buffy patches to sides of face and neck.
Wings Brown, streaked black with a white wingbar. Underparts Grey.

Conspicuous feature Whitish wingbar.

Conspicuous characteristics Tends to flock in large numbers on pasture · Enjoys roadside verges · Returns to noisy evening roosts outside of the breeding season · A late nester, breeding from early October, when it can be seen carrying streamers of straw into the treetops or under house eaves · The female can be confused with other finches, especially the Redpoll.

Call A noisy 'chirrup' and chattering sound is the usual song. The House Sparrow dawn chorus starts after the Blackbird and Song Thrush have finished singing.

Nest A bundle of straw in trees or buildings with an entrance hole. Four white, brown-spotted and streaked eggs are laid.

Target localities Gardens and parks.

Indian Myna (mature)

Conspicuous colours Brown, white and black.
Habitat A bird of suburban gardens, parks, open country, seldom venturing into forests, although occasionally seen along forest verges.
Range North Island only. Largest populations are to the north of Lake Taupo but scattered populations are found south to Wellington.
Size 240 mm (House Sparrow 145 mm).
Abundance Common.
Status Introduced.

Family Sturnidae
Species *Acridotheres tristis*
Common name Myna
Description – immature
Head Dark Brown.
Upperparts Brown.
Underparts Light-brown to white undertail.
Description – mature
Head and neck Glossy black.
Upperparts Cinnamon-brown.
Wings Cinnamon-brown with black primaries and prominent white patches.
Tail Black with a white tip.
Underparts Light brown.
Underwing and tail White.
Bill, legs and feet Yellow.
Eye Bare patch behind eye is yellow.

Conspicuous features White patches on wings when bird flies ✎ White tail tip ✎ Yellow bill.

Conspicuous characteristics Has a habit of feeding on tarmac roads in late afternoon ✎ Usually seen flying in pairs or in groups of up to ten.

Call Song delivered by both sexes, is a loud bell-like assemblage of notes which is more conspicuous in the autumn months when most of the European songsters have stopped singing.

Nest A bundle of grass and sticks in a hole in a bank or in an old building. Up to five greenish-blue eggs are laid.

Target localities Common in most places in the North Island, in towns to open country.

Canada Goose

Conspicuous colours Brown, white and black.
Habitat Rivers, lakes and back country pasture.
Range Throughout New Zealand except the Far North.
Size 830 mm (House Sparrow 145 mm, Feral Goose 800 mm).
Abundance Common.
Status Introduced.

Family Anatidae **Species** *Branta canadensis*

Common name Canada Goose

Description Head and neck Black. Face Prominent white patch under eye to chin. Upperparts Brownish. Breast and underparts Greyish-brown barred with white. Uppertail Black with white under. Bill Black. Legs and feet Dark grey.

Conspicuous features Contrasting black and white on the head ✎ Greyish-brown breast barring.

Conspicuous characteristics Easily scared and quickly takes to the air if disturbed ✎ Rests on the water in large groups away from disturbance ✎ In-flight birds fly in scattered skeins.

Call A musical double honk.

Nest A solitary nester which makes a bowl of twigs and grass, lined with feathers, usually on a mound with good visibility. Up to five white eggs are laid.

Target localities Lake Pupuke, Takapuna, Auckland ✎ Lake Rotoroa or Hamilton Lake, Hamilton ✎ St Anne's Lagoon, Cheviot, North Canterbury ✎ Lake Dunstan, Cromwell.

Spur-winged Plover

Conspicuous colours Brown, white and black.
Habitat Suburbs, parks, open country, coast.
Range Found throughout New Zealand except for Fiordland.
Size 380 mm (House Sparrow 145 mm).
Abundance Common.
Status Native. An arrival from Australia that first established breeding populations in Southland in the 1950s.

Family Charadriidae

Species *Vanellus miles*

Common names Spur-winged Plover, Masked Plover (Australia)

Subspecies Two are recognised: the northern Australian form (*V. m. miles*), and the southwestern and southeastern Australian form (*V. m. novaehollandiae*). The New Zealand population stems from *novaehollandiae*. This is distinctive in having a black crown with black markings extending down the neck to the mantle and to the shoulder. *Miles* has a black 'skull cap' only.

Description Head Crown and neck to mantle and shoulders black. Face Yellow facial wattle to behind eye. Wings Brown with black primaries. Rump White. Tail White with a black tip. Underparts White. Bill Yellow. Legs Reddish.

Conspicuous features Bony wing spur on carpal flexure joint of each wing ✎ Yellow mask ✎ In-flight birds show black primary feathers on wings ✎ In-flight birds show a black tail-band with a white rump and tail-tip.

Conspicuous characteristics When resting it often stands with head held high and back sloping ✎ Runs and pauses when feeding ✎ Wings look short and rounded when in flight ✎ Has slow wing-beats but with a faster down-beat ✎ Its high-pitched call gives it away.

Call A high-pitched agitated 'kitter kitter kitter'. Calls day and night.

Nest A scrape on pebbles or pasture with minimal nesting material. Up to five khaki, dark-brown-blotched eggs are laid. Main nesting season is from August through until April.

Target localities Parks near rivers ✎ Miranda, Firth of Thames ✎ Invercargill to Te Anau.

Black-fronted Dotterel (breeding)

Conspicuous colours Brown, white and black.
Habitat Coast, rivers and estuaries.
Range Selected localities throughout New Zealand.
Size 170 mm (House Sparrow 145 mm, Wrybill 200 mm).
Abundance Common.
Status Native.

Family Charadriidae **Species** *Charadrius melanops*

Common name Black-fronted Dotterel

Description Crown Mottled light brown. Forehead and face A black bar passes from the bill through the eye, to back of head. Eyebrow White. Upperparts Light brown mottled with darker tonings. Chin and throat White, beneath which is a V-shaped black band extending to back of neck. Underparts White. Bill Red with black tip. Legs Dull pink.

Conspicuous features Colourful with contrasting colours of brown, black and white ✎ Black forehead and white eyebrow ✎ Deep black V breast-band beneath white throat.

Conspicuous characteristics Blends with the river stones ✎ Always found feeding close to water ✎ Flight is jerky and dipping.

Call A fast-clicking 'tik-tik-tik-tik'.

Nest A scrape on sand or among river pebbles, lined with fine grass. Up to three eggs, khaki in colour and marked with brown spots and lines are laid.

Target localities Puketutu Island near Mangere sewage plant, Auckland ✎ Matata Lagoons, Bay of Plenty ✎ Tukituki River, Hawke's Bay.

Shore Plover (mature)

Conspicuous colours Brown, white and black.
Habitat Coast and estuaries.
Range Chatham Islands, but a few have been transferred to Portland Island off the Mahia Peninsula north of Napier.
Size 200 mm (House Sparrow 145 mm, Wrybill 200 mm).
Abundance Rare.
Status Endemic.

Family Charadriidae **Species** *Thinornis novaeseelandiae*

Common names Shore Plover, Tuturuatu

Description – breeding male Crown Greyish-brown above a white head-ring. Face and neck Black from above eye to throat. Upperparts Greyish-brown flecked with black. Underparts White. Bill Red tipped with black. Legs Orange.

Description – breeding female Upperparts and underparts Paler than the male bird.

Description – immature Upperparts Brown. Eyebrow stripe White. Underparts White.

Conspicuous characteristics Runs with short, fast steps before pausing A sedentary species that stays on its same area of beach throughout the year.

Call A high-pitched 'peep pee'.

Nest Usually hidden on the ground under vegetation. Up to three buff, brown-blotched eggs are laid.

Target locality Might turn up on mainland from time to time in Gisborne or Napier areas.

Turnstone

Conspicuous colours Brown, white and black.
Habitat Coast and estuaries.
Range Scattered throughout New Zealand.
Size 230 mm (House Sparrow 145 mm, Wrybill 200 mm).
Abundance Uncommon.
Status Migrant.

Family Charadriidae **Species** *Arenaria interpres*

Common names Turnstone, Ruddy Turnstone

Description – breeding Crown Brown, lightly flecked with white. Face Black and white. Chin White. Neck and breast Black and white. Wings Brown inner leading edge, black outer leading edge, with white primaries. Abdomen Clean white. Uppertail Black, then band of white and a broad, black tip. Bill Black. Legs Orange.

Description – non-breeding Upperparts and underparts A dull version of the breeding plumage.

Conspicuous features Black and white colours around head and breast • White upperwing primary feathers when bird is in flight • Uppertail black, then white and black tip.

Conspicuous characteristics Enjoys washing and preening in fresh water at rivermouths • At high-tide roosts, seldom mixes with other species • Busily fossicks when feeding, flipping stones.

Call A 'titititit' sound.

Target localities Clarks Beach, Manukau Harbour, west of Papakura • Miranda, Firth of Thames • Matata Lagoons, Bay of Plenty • Manawatu Estuary near Foxton • Motueka Estuary, Marlborough.

Arctic Skua (light phase)

Conspicuous colours Brown, white and yellow.
Habitat Sea.
Range Around New Zealand.
Size 430 mm (House Sparrow 145 mm, Pomarine Skua 480 mm).
Abundance Uncommon.
Status Migrant

Family Stercorariidae

Species *Stercorarius parasiticus*

Common names Arctic Skua, Arctic Jaeger

Phases The Arctic Skua comes in both a dark and a light phase. They are known for a variation of colour so the descriptions below are approximate only.

Description – breeding (light phase) Crown Black. Nape Pale yellow. Upperparts Brown. Chin and throat White. Breast Light-brown band. Underparts White. Tail Brown, with a protruding central tail feather.

Description – non-breeding (light phase) Crown, neck and upperparts Dark brown. Chin and throat White. Breast Brown band. Underparts White. Tail Brown with a protruding central tail feather.

Description – non-breeding (dark phase) Upperparts and underparts Brown with crown darker. Wings Brown with conspicuous white-flashes at the leading edge of the primary wing feathers on the underwings and with traces of white on the upperwings. The wing-flashes are sometimes missing.

Conspicuous features Prominent tail feather ✎ White underwing flashes.

Conspicuous characteristics Preys on fish caught by terns, although it also catches its own food ✎ Flies steadily and level when searching for feeding terns ✎ Extremely manoeuvrable when in flight chasing terns.

Breeding localities Northern hemisphere north of the 58th parallel in loose colonies on offshore islands, open tundra and on open sandspits.

Migration Birds head southwards into the southern hemisphere in July to the 46th parallel. They return north in March. Some immatures are known to stay on.

Target localities Waters out from Sandspit ✎ Waters of the Hauraki Gulf ✎ Miranda, Firth of Thames.

Pomarine Skua (light phase)

Conspicuous colours Brown, white and yellow.
Habitat Sea.
Range Around New Zealand.
Size 480 mm (House Sparrow 145 mm, Arctic Skua 430 mm).
Abundance Uncommon.
Status Migrant.

Family Stercorariidae **Species** *Stercorarius pomarinus*

Common name Pomarine Skua

Phases The Pomarine Skua comes in both a dark and a light phase.

Description – breeding (light phase) Crown Black. Nape Pale yellow. Upperparts Dark brown. Chin and throat White. Breast Brown band. Underparts White. Tail Brow, with a protruding central tail feather.

Description – non-breeding (light phase) Crown, neck and upperparts Dark brown. Chin and throat Brown. Underparts Brown. Tail Brown.

Conspicuous features Prominent tail feather ✎ White underwing flashes ✎ White underparts.

Conspicuous characteristics Preys on fish caught by terns, although it also catches its own food ✎ Flies steadily and level when searching for feeding terns ✎ Extremely manoeuvrable when in flight chasing terns.

Breeding localities of the Pomarine Skua Northern hemisphere north of the 58th parallel in loose colonies on offshore islands, open tundra and on open sandspits.

Migration To the southern hemisphere in July south to the 46th parallel. They return north in March. Some immatures are known to stay on.

Target locality Waters out from Sandspit in late summer.

Chestnut-breasted Shelduck

Conspicuous colours Chestnut, white and black.
Habitat Pasture, lake edges and lakes.
Range Most records come from the South Island with isolated sightings also coming from South Auckland.
Abundance Rare.
Size 630 mm (House Sparrow 145 mm, Paradise Shelduck 630 mm).
Status Native.

Family Anatidae

Species *Tadorna tadornoides*

Common name Chestnut-breasted Shelduck

Note Has been regularly seen in New Zealand since 1982 when it was first recorded on Lake Ellesmere. It is known to have bred here twice.

Description – male Head and neck Black with a metallic green sheen and a white ring around the edge of the black at the base of the neck. Upperback and breast Orange-chestnut. Wings Black with prominent white wing-coverts on upper and underwings, and a large green speculum. Underparts Brown, finely barred with white. Bill Black with small white patch at base. Legs and feet Dark grey.

Description – female Similar to the male except for a white eye-ring and a larger amount of white around the base of the bill.

Conspicuous features Bright chestnut breast separates it from the male Paradise Shelduck. White neck-ring and white ring around bill separate it from the male Paradise Shelduck. In flight, birds show large areas of white on both upper and underwings.

Conspicuous characteristic Usually in pairs on pasture or near water. Sometimes in family groups.

Call Male bird makes a 'zizzing-zonk' sound while the female makes a high-pitched, two-syllabic 'ong-chank' sound. When put to flight both birds usually call in an agitated manner, not unlike the Paradise Shelduck.

Nest In a hole in a tree or stump. Up to nine white eggs are laid.

Moult In Australia birds congregate in large numbers on inland lakes where they seek the sanctuary of deep water. This habit is similar to the Paradise Shelduck.

Target localities Puketutu Island, Manukau Harbour, Auckland; Patetonga Lagoon on the Hauraki Plains; Lake Elterwater, south of Blenheim.

Paradise Shelduck (female)

Conspicuous colours Chestnut, white and black.
Habitat Pasture, ponds, shallow freshwater areas and lakes.
Range Throughout New Zealand.
Abundance Common.
Size 630 mm (House Sparrow 145 mm, Mallard 580 mm).
Status Endemic.

Family Anatidae

Species *Tadorna variegata*

Common names Paradise Shelduck, Paradise Duck or Parry, Putangitangi

Description – female Head White with black eye. Breast and underparts Orange-chestnut, tending to brownish when not in breeding plumage. Wings Black with prominent white wing-coverts on upper and underwings, and a large green speculum. Undertail Orange-chestnut. Bill and legs Black.

Description – male Head and neck Black, with a metallic green sheen. Upper and underparts Black, lightly barred with white. Wings Black, with prominent white wing-coverts on upper and underwings, and a large green speculum. Abdomen Reddish-brown. Undertail Orange-chestnut. Bill Black. Legs and feet Black.

Description – duckling Zebra-striped brown and white when first born.

Description – immature Fledglings of both sexes resemble the adult male.

Conspicuous features The white on the head of the female ✎ The white on the wings.

Conspicuous characteristics Usually seen in pairs ✎ After the breeding season family parties are encountered ✎ When circling after disturbance birds call in a duet manner ✎ Ducklings have a habit of sitting in little pyramid-like heaps, always within sight of the parent birds. If disturbed they quickly scatter.

Call Female – a high-pitched 'ziz zik'. Male – a deep 'klonk klonk'.

Nest In holes in the ground or in old rotting stumps or rock crevices but sometimes in holes in trees. Up to nine white eggs are laid.

Target localities Open countryside ✎ Patetonga Lagoon on the Hauraki Plains ✎ Lake Rotoehu near Rotorua in summer.

Pheasant (male)

Conspicuous colours Chestnut, white and black.
Habitat Pasture and wooded gullies.
Range North Island and Canterbury in the South Island.
Size 800 mm (House Sparrow 145 mm, California Quail 250 mm).
Abundance Common.
Status Introduced.

Family Phasianidae

Species *Phasianus colchicus*

Common name Pheasant

Subspecies Three subspecies of Pheasant were originally introduced to New Zealand for hunting purposes. They were the Black Pheasant (*P. c .colchicus*) from Europe, the White-collared Pheasant (*P. c. mongolicus*) from Mongolia and the Ring-necked Pheasant (*P. c. torquatus*) from China. Today the most common subspecies is the Ring-necked Pheasant from China. There is also the Melanistic Mutant Pheasant (*P. c. tenebrosus*), breeding true to form in some areas.

Description – (male) *P. c. torquatus* Face Red. Head Black with bluish tinge on neck and white neck-collar. Two short ear-tufts are on each side of the crown. Wings Light brown streaked with white and dark brown. Back Golden-brown with black speckling. Rump Golden-brown. Tail Buff, striped horizontally with up to 16 separate black bars. Underparts Golden-buff with some black speckling. Bill Light brown. Legs Grey.

Description – female for all subspecies Upper and underparts Overall a dull, buff-coloured bird.

Description – (male) *P. c. colchicum* Less colourful than *torquatus*, being darker and lacking the white neck-collar.

Description – (male) *P. c. mongolicus* Red face wattle larger, plumage darker, and neck-collar thinner than *torquatus*. Wings lack grey colouring.

Description – (male) *P. c. tenebrosus* Back, breast and underparts Purplish-blue scalloped with buff and red. Wings Olive-black. Tail Olive-brown with black horizontal barring.

Description – (female) *P. c. tenebrosus* Purplish-black.

Conspicuous features The short ear-tufts.

Conspicuous characteristics Often seen gliding across gullies When disturbed in grass will rise quickly with very fast wing-beats.

Call Loud 'kor-r-kok' crow, accompanied whirring of wings before and after calling.

Nest In dense grasses on the ground. Up to nine olive-brown eggs are laid.

Target localities Open pasture in North Island Cornwall Park, Auckland Miranda - Kaiaua, South Auckland for the Melanistic Mutant Pheasant (*P. c. tenebrosus*). Further north of Kaiaua and south of Miranda *P. c. torquatus* becomes the common variety.

Silvereye

Conspicuous colour Green.
Habitat Gardens, parks, forests.
Range Throughout New Zealand.
Size 120 mm (House Sparrow 145 mm).
Abundance Common.
Status Native.

Family Zosteropidae

Species *Zosterops lateralis*

Common names Silvereye, White-eye, Tauhou, Grey-backed Silvereye (Australia)

Description
Head, nape and rump Green. Back Grey. Wings Green, primaries edged with brown. Chin Yellowish. Underneck and breast Pale grey. Flanks Pinkish-brown. Abdomen Whitish. Eye-ring White. Tail Above green, edged with brown, with white undertail.

Conspicuous features Small, green bird ✎ White eye-ring ✎ Grey back.

Conspicuous characteristics Gathers into winter-feeding flocks which fossick through areas of vegetation in a noisy manner ✎ Often seen feeding with Grey Warblers in canopy trees. The white tips to the Grey Warbler's tail separates it at a distance.

Call Several Silvereye songs have been detected. The alarm note is a noisy 'tchirrup', while the territorial song, delivered by the male bird, is rapid and warbled, not unlike a thin rendition of a Blackbird's song. In autumn, birds often sing softly from within trees and shrubs.

Nest A rather flimsy cup woven from fine grasses, mosses and twigs, hung between two branches, and very often placed at the extremity of branches. Up to 3 m from the ground. Three pale blue eggs are laid.

Breeding habits Because Silvereyes often place their nests on the end of flimsy branches, it is likely they are beyond the reach of rats (*Rattus rattus*). It breeds from September through to March.

Target localities Parks and gardens.

Greenfinch

Conspicuous colour Green.
Habitat Gardens, parks, open country, forests.
Range Throughout New Zealand.
Size 120 mm (House Sparrow 145 mm).
Abundance Common.
Status Introduced.

Family Fringillidae

Species *Carduelis chloris*

Common name Greenfinch

Description – male (breeding)
Face Grey, a black marking extending from the upper bill to behind the eye. Eyebrows Bright yellow. Upperparts Olive-green. Wings Primaries are black, secondary feathers grey, with a bright-yellow bar in middle of wing. Underwing Grey and yellow. Breast Yellowish-green above, with a yellow patch on belly. Flanks Grey. Rump Yellowish. Tail Greyish inner feathers, yellowish outer feathers, with a dark tip. Eye Brown.

Description – female (non-breeding)
Upper and underparts Similar to the male bird but much duller.

Conspicuous features Dark eye • Green head and back • Yellow leading edge of wing of the standing bird.

Conspicuous characteristics Sings from high in conifers • Flocks on to mown grass in winter and spring with other finches • During the breeding season singing males will fly from perch to perch • Flight is bat-like with wings beaten in a staccato fashion • Tail is slightly forked and slightly splayed in flight.

Call Song, delivered by the male bird, is a drawn-out 'tuusweeet'. Also a melodious 'chichichichi - tuu tuu tu' warble. Songs are uttered from high perches, especially from conifer trees. Note that its song is often mistaken for the Long-tailed Cuckoo. The Greenfinch, though, has notes arranged on a descending scale while the cuckoo has notes arranged on an ascending scale. Song period starts in September and continues into February.

Nest A bulky cup-like structure of twigs, leaves and moss, lined with fine dry grass, wool or feathers, placed near the top of a bush, in a fork or on a branch. Up to five pale-blue, brown-blotched eggs are laid.

Target localities Parks, gardens and open country.

Shining Cuckoo

Conspicuous colour Green.
Habitat Gardens, parks, forests.
Range Throughout New Zealand.
Size 160 mm (House Sparrow 145 mm).
Abundance Common.
Status Native migrant that winters to the northwest of New Zealand from the Solomon Islands to eastern Indonesia. It breeds in Australia, New Caledonia, Vanuatu and New Zealand. It uses other birds (in New Zealand the Grey Warbler in particular) to incubate and rear its young.

Family Cuculidae

Species *Chrysococcyx lucidus*

Common names Shining Cuckoo, Pipiwharauroa

Description Upperparts Shining green. Underparts White, finely barred with metallic green.

Conspicuous features Green barring on breast A flash of green often gives the bird away when it is in flight.

Conspicuous characteristics Mating birds tend to congregate in tall trees, in particular eucalypts, where they carry out noisy behaviour. This often includes fighting between males and courtship feeding with moths between pairs The song gives the bird away.

Call An easily recognised 'tu-wee tu-wee tu-wee tu-wee tee-wuu' whistle. Upon arrival in New Zealand in the spring, birds are very noisy. Cuckoo song is seldom heard after the end of January. There is evidence to suggest that Shining Cuckoos, when in New Zealand, are in fact a territorial species, actually living throughout the breeding season within a defined area of the territories of several pairs of Grey Warblers. Cuckoo song might therefore proclaim a territory as well as attract a mate.

Nesting Shining Cuckoos do not make a nest or rear young. Instead they prey on the nests of other species. The preferred host is the Grey Warbler but Silvereyes and Fantails are also on record as having reared cuckoo chicks. The cuckoo lays one olive-green egg. How it is placed in the dome-shaped Grey Warbler's nest, with its small entrance hole on the side, which is too small for a cuckoo to enter, is unknown. It is possible though that the cuckoo breaks its way in, after which the Grey Warblers then repair the damage.

Date of arrival In northern New Zealand from the first week in September.

Date of departure Early February but some birds are still here at the end of March. These may be young birds of late broods.

Target localities Cornwall Park, Auckland Gardens, parks and forested areas.

Bellbird (male)

Conspicuous colour Green.
Habitat Gardens, parks, forests.
Range Absent from Waikato, Auckland and Northland, and in some areas of South Canterbury and Otago.
Size 200 mm (House Sparrow 145 mm).
Abundance Common in selected areas.
Status Endemic.

Family Meliphagidae

Species *Anthornis melanura*

Common names Bellbird, Korimako

Description – male
Head Olive-green with purple iridescence on forehead and crown. Upperparts Olive-green. Wings Dark bluish-black with yellow at bend of folded wing. Tail Bluish-black and slightly forked. Underparts Pale green. Eye Red. Bill Black.

Description – female Upperparts Tending to olive-brown. Cheek Whitish stripe under eye from gape. Wings and tail Brownish-black, tail slightly forked. Underparts Pale green. Bill Black.

Conspicuous features Green of the male bird · Purple head tonings on the male bird · Curved honeyeater bill · Female bird is duller than the male · Slightly forked tail.

Conspicuous characteristics Renowned for its early-morning chorus · Takes nectar from flowers in an acrobatic manner · Flight is fast with noisy wing rustle · In the forest it flies low and is manoeuvrable. In the open its flight becomes direct with some undulation.

Call Bell-like, liquid, clear and melodic which starts well before sunrise. Both sexes sing, male birds being of stronger voice and more repetitive. Also has a 'pek pek pek' alarm call.

Nest A loosely built structure of twigs and fern leaves, lined with fine grasses, usually in a fork of a tree, at about 4 m from the ground. Up to four white, brown-blotched eggs are laid.

Target localities Tiritiri Matangi Island, near Whangaparaoa · Opepe Forest near Taupo · Mt Peel Forest, Canterbury · Lake Gunn, Eglinton Valley.

Yellow-crowned Parakeet

Conspicuous colour Green.
Habitat Forests.
Range Heavily forested areas of the far north, and central and southern North Island. Forested areas of the South Island.
Size 250 mm (House Sparrow 145 mm, Red-crowned Parakeet 280 mm).
Abundance Common in selected areas.
Status Endemic.

Family Psittacidae

Species
Cyanoramphus auriceps

Common names
Yellow-crowned Parakeet, Kakariki

Description
Forehead Yellow to above the eye. Upperparts Green. Wings Green, with violet-blue on the primaries. Underparts Yellowish-green.

Conspicuous features
Smaller than the Red-crowned Parakeet ✎ Yellow crown on head.

Conspicuous characteristics
Rattled song in treetops ✎ Usually in the heavy upper branches of the forest ✎ Long tail and fast flying ✎ Occupies a higher niche in the trees than the Red-crowned Parakeet, which often feeds on or near the ground.

Call A constant chatter of 'chit chit chit' notes, but when alarmed, such as when a Long-tailed Cuckoo strays into its territory, notes become rattled and agitated.

Nest In holes in old trees up to 6 m above ground. Up to seven white eggs are laid.

Target localities Pureora Forest near Benneydale, King Country ✎ Waikareiti track near Lake Waikaremoana ✎ Lake Rotopounamu near Tokaanu ✎ Lake Gunn, Eglinton Valley ✎ Ulva Island, Stewart Island.

Orange-fronted Parakeet

Conspicuous colour Green.
Habitat Forests.
Range Hawdon Valley near Arthur's Pass and Hurunui Valley near Lake Sumner, North Canterbury.
Size 280 mm (House Sparrow 145 mm, Red-crowned Parakeet 280 mm).
Abundance Rare in selected areas.
Status Endemic. Confirmed as a separate species after DNA testing and habitat observation. Originally it was grouped with the Yellow-crowned Parakeet (*Cyanoramphus auriceps*), following crossbreeding experiments carried out in captivity. It is now known to be a close relative of the Red-crowned Parakeet.

Family Psittacidae
Species
Cyanoramphus malherbi
Common name
Orange-fronted Parakeet
Description
Forehead Orange to above the eye. Upperparts Pale green. Wings Green with violet-blue on the primaries. Underparts Yellowish-green.
Conspicuous features
Orange crown on head. Pale-green colouring.
Conspicuous characteristics
Rattled song. Occupies a canopy niche in the forest.
Call 'Chit chit chit' sounds.
Nest In holes, high up, similar to Yellow-crowned Parakeet.
Target area Hawdon Valley, Arthur's Pass.

Red-crowned Parakeet

Conspicuous colour Green.
Habitat Forests, open areas.
Range Offshore islands to the north of the North Island, the Urewera and Raukumara forests and on Stewart Island in the south.
Size 280 mm (House Sparrow 145 mm, Yellow-crowned Parakeet 250 mm).
Abundance Common in selected areas.
Status Endemic.

Family Psittacidae **Species** *Cyanoramphus novaezelandiae*

Common names Red-crowned Parakeet, Kakariki

Description Forehead A red cap extends from bill to above the eye. Upperparts Green. Wings Green with violet-blue primaries. Underparts Yellowish-green.

Conspicuous features Green colouring ✎ Long tail for its size.

Conspicuous characteristics Chattering call ✎ Fast flight ✎ Seeks the security of the inside branches of a tree ✎ Often feeds on the ground ✎ Enjoys feeding on flax seeds.

Call A rapid series of 'chit chit chit chit' notes made when in flight.

Nest A hole in trees about 5 m up, but also low to the ground in flax bushes, rock crevices and undergrowth. Up to seven white eggs are laid.

Target localities Tiritiri Matangi Island near Whangaparaoa ✎ Ulva Island, Stewart Island.

Kea

Conspicuous colour Green.
Habitat Forests, open country and high country.
Range South Island only.
Size 460 mm (House Sparrow 145 mm, Red-crowned Parakeet 280 mm).
Abundance Common in selected alpine areas. For some years they were blamed for killing sheep in the high country and were controlled. Today they are protected and numbers have increased.
Status Endemic.

Family Psittacidae **Species** *Nestor notabilis* **Common name** Kea

Description Upperparts and underparts Olive-green. Underwing Scarlet. Rump Dull red. Tail Has a touch of blue on longer feathers. Otherwise green.

Conspicuous features A large, green parrot ✎ Sharp-hooked bill ✎ Yellow flecking on body feathers ✎ Red underwing.

Conspicuous characteristics Confident when on the ground ✎ Calls noisily when in flight ✎ Will land on buildings and cars ✎ Comical like the Kaka ✎ Destructive to car windscreen wipers, tyre rims and valve stems.

Call The Kea screech is commonly heard in many alpine regions. It sounds like its name 'kee-aaa kee-aaa kee-aa'. Usually it comes from a perch above. It is also uttered when birds are in flight.

Nest In holes near the ground or under fallen logs. Up to four white eggs are laid.

Target localities Lake Rotoiti, Nelson Lakes National Park ✎ Alpine areas near Arthur's Pass ✎ Franz Josef, Westland ✎ Lake Gunn, Eglinton Valley ✎ Homer Tunnel, Milford Road, Fiordland.

Mallard (male)

Conspicuous colour Green.
Habitat Found in both fresh and saltwater environments.
Range Found throughout New Zealand and on offshore islands.
Size 580 mm (House Sparrow 145 mm, Grey Duck 550 mm).
Abundance Common.
Status Introduced.

Family Anatidae **Species** *Anas platyrhynchos* **Common name** Mallard

Description – male Head Dark, glossy green. Breast Chestnut. Upper and underparts Grey. Wings Grey. Speculum Blue. Bill, legs and feet Orange.

Description – female Head Brown with dark-brown eye-stripe. Upper and underparts Streaked and spotted brown and buff. Wings Brown with blue speculum. Bill Orange-brown.

Description – eclipse bird Upper and underparts Male birds attain feather patterns similar to female birds.

Conspicuous features Female Mallards, and hybrids between Mallards and Grey Ducks, lack the yellow and black face markings of Grey Ducks. Mallards appear lighter in eclipse plumage than Grey Ducks.

Conspicuous characteristics Found near human habitation such as on park lakes. Mallards usually take to the air quicker than Grey Ducks when disturbed. Head bobbing, neck stretching and circling of the male by the female on the water is noticed as birds pair-bond after February.

Call The male makes a high-pitched 'quek quek' call or just a piping whistle. The female makes a raucous quack.

Nest In grass or rushes, a bowl lined with moss. Up to 12 greenish eggs are laid.

Target localities Parks, public gardens and lakes.

Kakapo

Conspicuous colour Green.
Habitat Forests.
Range Confined to islands off Fiordland and Stewart Island.
Size 630 mm (House Sparrow 145 mm, Red-crowned Parakeet 280 mm).
Abundance Rare.
Status Endemic. An ancient New Zealand species probably related to the Australian Night Parrot (*Pezoporus occidentalis*) and the Ground Parrot (*Pezoporus wallicus*). In predator-free New Zealand it has become large, slow-breeding and flightless. It is a nocturnal species that feeds and carries out its courtship behaviour at night. It now totals around 80 birds.

Family Psittacidae **Species** *Strigops habroptilus* **Common name** Kakapo

Description Upperparts Green mottled with brown. Underparts Greenish mottled with yellow. Tail Brownish.

Conspicuous features Green all over A very large parrot.

Conspicuous characteristics Flightless Nocturnal feeding habits Climbs trees for some foods.

Call A booming noise is made by male birds at the breeding season when they call for a mate from booming bowls.

Nest On the ground in a clump of rushes or under a log. Up to four white eggs are laid.

Target locality Birds cannot be viewed as they live on inaccessible offshore islands where entry is prohibited.

Rifleman (male)

Conspicuous colours Green and brown.
Habitat Forests.
Range Found through central and lower North Island and the South Island excluding the Canterbury Plains and North Canterbury pasture. Also on Stewart Island.
Size 80 mm (House Sparrow 145 mm, Grey Warbler 100 mm).
Abundance Common in selected areas.
Status Endemic. The ancient family of Acanthisittidae is a good example of adaptive radiation, the original ancestor separating into the Rifleman (*Acanthisitta chloris*), a middle and upper forest dweller, the Bush Wren (*Xenicus longipes*), a bird of the forest floor and understorey, the Rock Wren (*X. gilviventris*), a bird of alpine rock garden areas, and the Stephen's Island Wren (*X. lyalli*). The Stephen's Island Wren and the Bush Wren are now both extinct.

Family Acanthisittidae

Species *Acanthisitta chloris*

Common names Rifleman, Titipounamu

Subspecies Two are recognised: North Island Rifleman (*A. c. granti*) and South Island Rifleman (*A. c. chloris*).

Description – male Head Bright yellow-green with white eye-stripe. Upperparts Bright yellow-green. Wings Green, with a noticeable yellow bar. Underparts Whitish. Tail Brown, and white-tipped.

Description – female Head Brown, slightly zebra-striped with white. Eyebrow White. Upperparts Brown, slightly zebra-striped with white. Wings Brown, with faint yellow bars on the secondary feathers and faint white markings on the inner edges of primary feathers. Underparts Whitish. Tail Brown, and white-tipped.

Conspicuous features The smallest bush bird ✎ Smaller than the Grey Warbler ✎ Appears tailless in comparison to the Grey Warbler ✎ White eyebrow on both male and female birds ✎ White tip to tail ✎ Pale yellow wingbars.

Conspicuous characteristics Has a habit of working up tree trunks, hopping from one side to the other ✎ Sometimes it wing-flicks as it feeds.

Call A high-pitched 'zit zit zit zit', which is difficult to hear.

Nest A hole or cavity in an old tree. Up to four white eggs are laid.

Target localities Pureora Forest, Benneydale, King Country ✎ Opepe Reserve, State Highway 5, Taupo ✎ Lake Rotopounamu, near Tokaanu ✎ Lake Waikareiti track near Lake Waikaremoana ✎ Pelorus Bridge, Marlborough ✎ Hinewai Reserve, Akaroa ✎ Lake Gunn, Fiordland.

Rock Wren (male)

Conspicuous colours Green and brown.
Habitat Alpine areas above the tree line, in rock gardens of scattered low vegetation along either side of the Southern Alps.
Range South Island only. Populations are higher in the Fiordland area.
Size 100 mm (House Sparrow 145 mm, Grey Warbler 100 mm).
Abundance Uncommon in selected areas.
Status Endemic.

Family Acanthisittidae

Species *Xenicus gilviventris*

Common names Rock Wren, Hurupounamu

Description – male
Head Dull green.
Eyebrow White edged with black.
Upperparts Dull green.
Underparts Grey-brown.
Flanks Yellow and green.
Bill Black and fine.

Description – female
Upperparts Olive-brown.
Underparts Grey-brown.
Flanks Yellow and green but paler than male.

Conspicuous features
Larger than the Rifleman ✎ White eyebrow ✎ Large toes, with prominent hind-toe.

Conspicuous characteristics Continually bobs its body when standing ✎ Will suddenly appear ✎ Hops quickly from rock to rock with some use of wings.

Call A high-pitched three-syllable 'zee zit zit' sound, the first note being rather piercing. The call is not often heard but is distinctive.

Nest A hole in a bank or rock crevice on the ground. Up to three creamy eggs are laid.

Target localities Arthur's Pass National Park ✎ Gertrude Valley, Hollyford, Fiordland ✎ Homer Tunnel, Milford Road, Fiordland ✎ Near the start of the Routeburn Track, Glenorchy, near Queenstown.

New Zealand Pigeon

Conspicuous colours Green and white.
Habitat A bird of both secondary and old forest, in particular of those which contain fruiting varieties such as puriri, kohekohe, taraire, tawa and miro. It also enters suburban gardens to feed on flowering or fruiting trees.
Range Found throughout forested areas of New Zealand and on the off-shore islands. It is a very common bird on Stewart Island.
Size 510 mm (House sparrow 145 mm, Rock Pigeon 330 mm).
Abundance Common in selected areas.
Status Endemic.

Family Columbidae

Species *Hemiphaga novaeseelandiae*

Common names New Zealand Pigeon, Kereru, Kukupa (Northland)

Description Upperparts Green with feathers showing a metallic purplish iridescence, especially around the neck. Underparts White, this white appearing like a white 'apron' on abdomen.

Conspicuous feature The white abdomen 'apron' on the roosting pigeon is conspicuous.

Conspicuous characteristics Pigeon 'swish' flight noises are audible ✎ Open-country flight is level and steady ✎ Courtship display habit of soaring out from high trees, gaining altitude and then suddenly stalling and swooping earthward is often seen.

Call Usually a soft 'kuu kuu kuu' sound is heard in the breeding season.

Nest A loose structure of sticks at about 4 m above the ground. One white egg is laid.

Target localities Wenderholm Park at Waiwera north of Auckland ✎ Tiritiri Matangi Island ✎ Pureora in the King Country ✎ Lake Gunn, Fiordland ✎ Half Moon Bay, Stewart Island.

Grey Warbler

Conspicuous colours Grey and white.
Habitat Parks, gardens, scrubland remnants, forests.
Range Throughout New Zealand.
Size 100 mm (House Sparrow 145 mm).
Abundance Common.
Status Endemic. Closely related to the Australian *Gerygone* genus and to the New Caledonian Yellow-sided Warbler (*Gerygone flavolateralis*).

Family Acanthizidae

Species *Gerygone igata*

Common names Grey Warbler, Riroriro

Description

Upperparts Grey-brown.
Face, throat and breast Pale grey.
Underparts White.
Tail Grey with white tips, the white being absent on the central tail feathers.

Conspicuous features Long tail with white tips to tail feathers ⟋ Smaller than a House Sparrow but larger than a Rifleman ⟋ The long tail separates the bird from a Rifleman.

Conspicuous characteristics
Often feeds on the wing, fluttering as if trying to fan leaves to dislodge insects ⟋ Moves quickly through the branches, usually in the middle canopy ⟋ Regularly warbles as it feeds ⟋ Often flies high above the canopy, chasing other birds in territorial battles.

Call Song is a sweet and fragile warble. Dialects vary from district to district although the bird is always identifiable. Birds quiver right to tip of tail when in full song. Birds can be heard singing at any time of year but they are particularly vocal throughout the spring and summer breeding season.

Nest Nests are of domed construction, pear-shaped, with a small hole on the side for an entrance. They are made of twigs, lichen, bark and moss and lined with feathers. Up to four white, brown-speckled eggs are laid.

Cuckoo parasitism The Grey Warbler is the major foster parent for young Shining Cuckoos. Therefore they nest early so that one brood of warblers is on the wing before the arrival of the cuckoo in September or October. In northern New Zealand the first brood is usually fully fledged by early August.

Target localities Common in many gardens and parks.

New Zealand Robin (female)

Conspicuous colours Grey and white.
Habitat Native forests but seldom found in small remnants. Also found in the exotic pine forests of the central North Island.
Range In the North Island confined to central areas and Mokoia Island in Lake Rotorua. Also on Great Barrier Island, Little Barrier Island, Tiritiri Matangi Island and Kapiti Island. In the South Island absent from Canterbury Plains and Central Otago. Found on Stewart Island and Ulva Island.
Size 180 mm (House Sparrow 145 mm, Tomtit 130 mm).
Abundance Common in selected areas.
Status Endemic.

Family Eopsaltriidae

Species *Petroica australis*

Common names New Zealand Robin, Totouwai

Subspecies Three, all black in colour: North Island Robin (*P. a. longipes*), South Island Robin (*P. a. australis*), Stewart Island Robin (*P. a. rakiura*).

Description – New Zealand Robin (female) Upper and underparts Dark grey. Lower breast and abdomen Dull white.

Description – North Island Robin (male) Upperparts Black. Frontal dot above bill White. Upper breast Almost black. Lower breast and abdomen White.

Description – South Island Robin (male) Upperparts Black. Frontal dot above bill White. Breast and abdomen White with hint of yellow.

Conspicuous features Larger than the Tomtit / Bold eye.

Conspicuous characteristics Has a confiding nature and will approach humans / Has a habit of just appearing / Scratches through leaf litter on the forest floor like a Blackbird.

Call The song, delivered by the male bird, is a ringing 'tueet tueet tueet tueet tooo' on a descending scale. Birds are more vocal in mid-morning. They also have a distinctive alarm call.

Nest A bulky cup of twigs, fern and moss bound with cobwebs and lined with feathers, wool or tree-fern hairs. Up to four cream, brown-speckled eggs are laid.

Target localities Tiritiri Matangi Island near Whangaparaoa / Wenderholm Regional Park, north of Waiwera, North Auckland / Eglinton Valley near Te Anau / Ulva Island, Stewart Island.

Tufted Guineafowl

Conspicuous colours Grey and white.
Habitat Open pasture and rough areas of scattered vegetation.
Range On farmland mostly in Northland, Rotorua, Hawke's Bay, Wanganui and North Canterbury.
Size 600 mm (House Sparrow 145 mm, Wild Turkey 1200 mm).
Abundance Uncommon.
Status Introduced.

Family Phasianidae Species *Numida meleagris*

Common name Tufted Gunieafowl

Description Face Blue-grey bare skin with red wattles below bill.
Upper and underparts Grey spotted with white. Bill Tan.

Conspicuous features Bare blue-grey face / Red wattles.

Conspicuous characteristics Usually in flocks / Often noisy.

Call A raucous cackle.

Nest A cup-shaped bowl among grasses. Up to twelve pinkish-buff eggs are laid.

Target localities Isolated rural areas usually near farm dwellings.

Wrybill (breeding)

Conspicuous colours Grey and white.
Habitat Coast, estuaries and mudflats in the North Island. Braided rivers in the high country of the South Island.
Range Harbours of Northland, Auckland, Coromandel, Bay of Plenty, Hawke's Bay and Taranaki to Waikanae. South Island high country from North Canterbury to Otago.
Size 200 mm (House Sparrow 145 mm, Banded Dotterel 200 mm).
Abundance Common in selected areas.
Status Endemic.

Family Charadriidae Species *Anarhynchus frontalis*

Common names Wrybill, Ngutuparore

Description – breeding Forehead White with white eyebrow to behind eye. Upperparts Grey. Throat A black band crosses the throat when in breeding plumage. Underparts White. Bill Black. Legs Grey-green.

Differences between sexes Male birds have a black spot between the white frontal stripe and the grey of crown when in breeding plumage ✎ Male birds have a broader black breast-band than females.

Conspicuous features Black breast-band noticeable but indistinct on wintering birds ✎ Bill curved to the right, the only bird in the world with such a bill. A study (Pierce, 1979) concluded 'the bill appears to be pre-adapted for obtaining mayfly and caddisfly from their inactive diurnal positions on the undersurface of submerged stones, where they are not normally visible to Wrybills.'

Conspicuous characteristics Birds when feeding run and pause ✎ Has a habit of running and suddenly changing direction while feeding ✎ Runs in a hunched position with head pulled back into shoulders ✎ A confiding species ✎ Birds flock during winter at high-tide roosts.

Call A shrill 'weet weet' sound.

Nest A scrape among the pebbles of a South Island riverbed. Two pale grey, dark-brown-blotched eggs are laid.

Breeding grounds During the breeding season of late July to December birds are on the wide riverbeds of the Mid-Canterbury area. Nests are placed up to 200 m apart and territories are defended. Birds return to their wintering grounds of the North Island from the end of January onwards.

Migrants attracted to high-tide Wrybill roosts The following rare species will sometimes roost with Wrybills – Mongolian Dotterel, Large Sand Dotterel, Siberian Tattler, Terek Sandpiper, Turnstone, Sharp-tailed Sandpiper, Curlew Sandpiper and Red-necked Stint.

Target localities Miranda, Firth of Thames ✎ Cass River near Tekapo.

Wrybill (non-breeding)

Conspicuous colours Grey and white.
Habitat Outside of the breeding season muddy estuaries and inlets.
Range Outside of breeding season around the coast in Northland, Coromandel, Bay of Plenty, southern Taranaki and Hawke's Bay. Main concentrations are around Kaipara Harbour, Manukau Harbour, Firth of Thames and Tauranga Harbour.
Size 200 mm (House Sparrow 145 mm, Banded Dotterel 200 mm).
Abundance Common in selected areas.
Status Endemic.

Family Charadriidae

Species
Anarhynchus frontalis

Common names
Wrybill, Ngutuparore

Description – non-breeding
Forehead White with a white eyebrow to behind eye.
Upperparts Grey.
Throat White and an absence of a black band. Underparts White.
Bill Black.
Legs Grey-green.

Conspicuous features
No black breast-band
Bill curved to the right.

Conspicuous characteristics
Runs and pauses when feeding. Runs and suddenly changes direction. Runs in a hunched position with head pulled back into shoulders.

Target locality Firth of Thames.

Non-breeding

Breeding

Differences between sexes in breeding plumage

- Male birds can be distinguished from female birds by the presence of a black spot between the white frontal stripe and the grey of crown.
- Male birds have a broader breast-band than females.

Marsh Sandpiper

Conspicuous colours Grey and white.
Habitat A bird of harbours and estuaries usually preferring still-water areas.
Range Birds turn up from time to time in all the main wading bird areas of New Zealand.
Size 220 mm (House Sparrow 145 mm, Wrybill 200 mm).
Abundance Rare.
Status Migrant.

Family Scolopacidae Species *Tringa stagnatilis*
Common name Marsh Sandpiper

Description – non-breeding Upperparts Pale grey with a dark edge to folded wings. Underparts White. Eyebrows White. Bill Black, slim and pointed. Legs Greenish.

Description – breeding Upperparts Grey, flecked with brown. Throat and breast Grey, lightly flecked with brown. Bill Black. Legs Yellowish.

Conspicuous features Stilt-like appearance but much smaller than a stilt ✐ Long slender bill.

Conspicuous characteristics At high tide it keeps feeding in shallow water while other species roost ✐ A fast flier ✐ When in flight it calls ✐ To separate from the Lesser Yellowlegs (*Tringa flavipes*), which also occasionally arrives in New Zealand at similar places as the Marsh Sandpiper, look for a white V marking on the back, seen when in flight.

Call Widely spaced 'twee twee twee' notes.

Breeding In Central Asia, migrating to Africa, southern Asia and Australia.

Target locality Miranda, Firth of Thames.

Siberian Tattler

Conspicuous colours Grey and white.
Habitat A bird of harbours and estuaries, usually preferring still water.
Range Birds turn up from time to time at all the main wading-bird areas of New Zealand.
Size 250 mm (House Sparrow 145 mm, Wandering Tattler 270 mm).
Abundance Rare.
Status Migrant.

Family Scolopacidae Species *Tringa brevipes*

Common names Siberian Tattler, Grey-tailed Tattler

Description – non-breeding Head Grey with white eyebrow and a dark stripe beneath eye. Upperparts Smooth grey. Chin White. Underparts Grey breast with white abdomen. Sides and flanks Some light, horizontal barring. Tail Grey with some white, indistinct cross-barring. Bill Black and heavy. Legs Yellow.

Description – breeding Upperparts Smooth grey. Underparts Heavily barred dark grey with white undertail.

Conspicuous features Long and heavy bill ✎ Has a squat posture ✎ Heavier than most shorebirds ✎ Yellow legs.

Conspicuous characteristics At high-tide roosts close to the mud line ✎ When alerted bobs head and dips tail ✎ Is easily alerted and put to flight ✎ Usually a solitary species.

Call A 'two-eet two-eet', the second note being a higher pitch than the first.

Target locality Miranda, Firth of Thames ✎ Maketu Harbour, Bay of Plenty ✎ Matata Lagoons, Bay of Plenty.

Wandering Tattler

Conspicuous colours Grey and white.
Habitat A bird of harbours and estuaries, usually preferring still water.
Range Birds turn up from time to time in all the main wading-bird areas of New Zealand.
Size 270 mm (House Sparrow 145 mm, Wandering Tattler 270 mm).
Abundance Rare.
Status Migrant.

Family Scolopacidae

Species *Tringa incana*

Common name
Wandering Tattler

Description – non-breeding
Head Grey with white eyebrow and a dark stripe beneath eye. Upperparts Dark grey. Chin White. Underparts Grey breast with some white on abdomen. Sides and flanks Light, horizontal barring. Tail Grey with some white, indistinct cross-barring. Bill Black and heavy. Legs Yellow.

Description – breeding Upperparts Smooth grey. Underparts Heavily barred dark grey with white undertail, similar to the Siberian Tattler.

Call A musical 'tew tew tew tew tew tew teeew'.

Target locality Miranda, Firth of Thames.

Differences between Siberian Tattler and Wandering Tattler

- Wandering Tattler is bigger than the Siberian Tattler.
- In the hand, the length of the nasal groove can be seen to be longer on the Wandering Tattler, about double that of the Siberian.
- In breeding plumage the Wandering Tattler has undertail barring and barring on the abdomen and lower breast. The Siberian has indistinct barring undertail and lighter barring on breast and abdomen. Birds seen in New Zealand usually show only traces of barring.
- The Wandering Tattler has a very narrow eye-stripe behind the eye and less white in front of the eye than the Siberian Tattler.
- The Wandering Tattler has longer wings, making its tail appear shorter, and is an overall darker bird.
- The Wandering Tattler has a distinctive call. It is a musical trill of 'tew tew tew tew tew tew teeew' notes; the Siberian has a two-syllable 'two-eet two-eet', the second note being higher pitched.

Lesser Knot (non-breeding)

Conspicuous colours Grey and white.
Habitat Harbours and estuaries.
Range Throughout New Zealand.
Size 250 mm (House Sparrow 145 mm, Wandering Tattler 270 mm).
Abundance Common.
Status Migrant.

Family Scolopacidae

Species *Calidris canutus*

Common names Lesser Knot, Red Knot, Huahou

Description – non-breeding plumage
Upperparts and wings Grey with thin, white wingbar. Underparts Greyish-white with some speckling on flanks. Rump Barred with white and grey. Bill Brown. Legs Greyish-brown.

Description – breeding plumage
Head, neck and breast Chestnu t. Upperparts A mixture of black, dark brown, grey and white. Abdomen Rufous.

Note: The white leg band indicates that this bird is part of a New Zealand research programme into migration patterns.

Conspicuous features Birds are considerably smaller than the godwits they roost among ✎ Birds appear drab and grey in non-breeding plumage ✎ In breeding plumage birds are highly coloured. Hence the name Red Knot. This colouring starts in late summer ✎ Heavier appearance than the other sandpipers they appear with ✎ Squat posture ✎ White wingbar when in flight.

Conspicuous characteristics Have a habit of intermingling with godwits on high-tide roosts. Often they can be found grouped right in the middle of a godwit mob.

Call Usually silent. Before migration often a two syllable 'poor-me poor-me' sound.

Target localities Clarks Beach, Manukau Harbour, out from Papakura ✎ Miranda, Firth of Thames ✎ Motueka Estuary, Nelson ✎ Awarua Bay, Invercargill.

Black-tailed Godwit (non-breeding)

Conspicuous colours Grey and white.
Habitat Coast and estuaries.
Range Scattered throughout New Zealand.
Size 390 mm (House Sparrow 145 mm, Wrybill 200 mm).
Abundance Rare.
Status Migrant.

A solitary Black-tailed Godwit surrounded by Bar-tailed Godwits

Family Scolopacidae Species *Limosa limosa*

Common names Black-tailed Godwit, Asiatic Black-tailed Godwit.

Description – non-breeding Upperparts Smooth grey. Upperwings Grey with black edges to the secondary feathers, under which is a white band that passes through the black primaries. Underwing White edged with black. Underparts Whitish. Rump White, the white extending down the tail to a broad, black tip. Bill Greyish with a black tip.

Description – breeding Head, neck and breast Chestnut. Eyebrow Whitish. Upperparts Brownish with individual feathers buff-edged. Abdomen White with some side barring with brown. Legs Lead-grey.

Conspicuous features Similar in size to a Bar-tailed Godwit White upperwing marking White tail and black tail-tip band Straight bill, not up-curved like the Bar-tailed Godwit Legs trail behind the bird in flight Legs longer than those of the Bar-tailed Godwit.

Conspicuous characteristics Is easily scared, quickly taking to the air when disturbed.

Call When put to flight a loud 'wicka wicka wicka' call is made.

Target locality Miranda, Firth of Thames.

Hudsonian Godwit

Conspicuous colours Grey and white.
Habitat Coast and estuaries.
Range Scattered throughout New Zealand.
Size 390 mm (House Sparrow 145 mm, Wrybill 200 mm).
Abundance Rare.
Status Migrant.

Family Scolopacidae

Species *Limosa haemastica*

Common name
Hudsonian Godwit

Description – non-breeding
Upperparts Smooth grey. Upperwings Grey, with black edges to the secondary feathers, under which is a white band that passes through the black primaries. Underwing White edged with black. Underparts Whitish. Rump White, the white extending down the tail to a broad, black tip. Bill Greyish with a black tip.

Description – breeding Head, neck, breast and abdomen Chestnut. Eyebrow Whitish. Upperparts Brownish, with individual feathers buff-edged. Legs Bluish-grey.

Quick observation points when watching godwits White on wings of flying birds (Black-tailed or Hudsonian) Dark armpits on flying birds or on wing-raised birds (Hudsonian) Straight bill (Black-tailed) or slightly curved (Hudsonian).

Call 'Toe-wit toe-wit toe-wit' sound is made.

Target locality Miranda, Firth of Thames.

Differences between Black-tailed Godwit and Hudsonian Godwit

- Black-tailed Godwit has a straight bill; Hudsonian has a slightly up-curved bill.
- Underwing of the Black-tailed Godwit is white, edged with black; underwing of Hudsonian is dark brown on the armpits, tips of primaries and secondaries, and has a white band in the centre of the underwing.
- Black-tailed Godwit has lead-grey-coloured legs; Hudsonian has bluish-grey legs.
- Black-tailed Godwit has less abdominal chestnut colouring in the breeding season with the chestnut not extending undertail.

Fairy Tern

Conspicuous colours Grey, white and black.
Habitat Coast, estuaries and harbours.
Range Local waters off Northland and the Kaipara Harbour.
Size 250 mm (House Sparrow 145 mm, Fluttering Shearwater 300 mm).
Abundance Rare.
Status Native.

Family Laridae Species *Sterna nereis*

Common names Fairy Tern (not to be confused with the White Tern, *Gygis alba*, a tropical species, which sometimes goes by the name of Fairy Tern), New Zealand Fairy Tern, Tara-iti.

Description – breeding Forehead White to above the eye. Crown, nape and eye surround Black. Back and wings Light grey, but primary feathers darker. Neck and underparts White. Tail White and forked. Bill Yellow-orange. Legs Orange.

Description – non-breeding Crown White of forehead extends to mid-crown.

Conspicuous features Yellow bill to the tip ✐ Black eye surround ✐ Forked tail.

Call Both Fairy and Little Terns make high-pitched 'cheet' or 'peep' calls.

Nest and chicks A solitary nester. It lays two eggs in the sand, relying on camouflage to protect the chicks from aerial and other predators. Chicks on hatching cryptically blend with the colour of the sand. By day 21 birds have taken on grey plumage barred with dark grey and have assumed a rusty-coloured head.

Target localities Mangawhai, Northland, near the heads ✐ Papakanui Spit on the South Head of Kaipara Harbour.

Differences between breeding Fairy Tern and Little Tern

- Fairy has a yellow bill; Little has a yellow bill black-tipped.
- Fairy has a black cap that extends around the eye in a soft S curve. This black does not extend right to the bill; Little has a neat black cap that extends around the eye in a sharp V then down to the bill.
- Fairy is paler grey on wings and back; Little has dark-grey primary wing feathers.
- Fairy has a rounded forehead; Little a swept-back forehead.

Differences between non-breeding Fairy Terns and Little Terns

- Fairy has more black on the crown; Little Tern has a swept-back crown with black towards the nape.

Differences between immature Fairy Terns and Little Terns

- Immature birds are inseparable.
- Birds of both species have black bills and indistinct white crowns mottled with black feather scalation.
- Fairy Terns have a higher crown.

Field characteristics that can help separate the two species

- Fairy Terns feed in company; Little Terns are solitary feeders.
- Fairy Terns dive boldly like Caspian Terns; Little Terns' dives are gentle splashes.
- Fairy Terns' flight is level and direct; Little Terns hover with fast wing-beats, with heads down.
- Fairy Terns tend to roost alone; Little Terns intermingle on high-tide roosts with waders.

Little Tern

Conspicuous colours Grey, white and black.
Habitat Coast, estuaries and harbours.
Range Throughout New Zealand spasmodically.
Size 250 mm (House Sparrow 145 mm, Fluttering Shearwater 300 mm).
Abundance Uncommon.
Status Migrant.

Family Laridae Species *Sterna albifrons*

Common names Little Tern, Eastern Little Tern

Description – breeding Forehead White. Eye-stripe White. Crown Black, a black line extending through eye to base of bill. Back Grey. Wings Grey with dark edges to primaries. Underparts White. Tail White and forked. Bill Yellow-orange with a black tip. Legs Yellow.

Description – non-breeding Forehead and crown White receding to black cap at back of head. Bill Pale yellow with dark tip.

Description – immature Shoulder Black. Bill Dark.

Conspicuous features Yellow-orange bill with black tip Partial black cap and the black line through eye to bill White line above eye.

Conspicuous characteristics Sits with roosting shorebirds at high tide, usually near the tide line Hovers in an agitated manner before diving Very dainty when in flight.

Call Urgent 'peet peet' sounds.

Nest Not known to have bred in New Zealand.

Target localities Miranda, Firth of Thames Maketu Harbour, Bay of Plenty.

Black-fronted Tern

Conspicuous colours Grey, white and black.
Habitat Rivers, pasture and coastal areas.
Range In summer over rivers and pasture of the South Island. In winter in estuaries and harbours of South Island, and on beaches from Wellington to the Bay of Plenty in the North Island.
Size 290 mm (House Sparrow 145 mm, Fluttering Shearwater 300 mm).
Abundance Uncommon.
Status Endemic.

Family Laridae Species *Sterna albostriata*

Common names Black-fronted Tern, Tarapiroe

Description – mature Crown Velvet-black. Under-eye White line. Upperparts Soft bluish-grey. Cheeks, throat and underparts Soft bluish-grey. Rump White with a grey forked tail. Bill and legs Orange.

Description – immature Forehead White. Crown Black streaking. Nape White. Wings and underparts Grey. Bill Brown.

Conspicuous features Orange bill and legs ✎ Black crown coming right to the bill ✎ Bluish-grey underparts.

Conspicuous characteristics Flits and darts when in flight ✎ Upon spotting prey, hovers with fast wing-beats ✎ Blends with riverbed stones on the ground.

Call A high-pitched 'kit kit kit'. It is a more brittle-sounding call than that of the White-fronted Tern.

Nest A scrape among the pebbles of the riverbeds in loose colonies. Up to two greyish, brown-blotched eggs are laid.

Target localities Waikanae Estuary in winter ✎ Rangitata River, Canterbury ✎ Eglinton Valley, Fiordland ✎ Waiau River, Southland.

Black-billed Gull

Conspicuous colours Grey, white and black.
Habitat Coast, rivers and pasture.
Range Selected areas in the North Island such as Manukau Harbour, Miranda and Rotorua. Throughout the South Island.
Size 370 mm (House Sparrow 145 mm, Red-billed Gull 370 mm).
Abundance Common.
Status Endemic.

Family Laridae **Species** *Larus bulleri*

Common names Black-billed Gull, Tarapunga

Description Head, neck and underparts Pure white. Wings Soft grey. Primaries Soft grey, edged with black and a trace of white. Bill and legs Black.

Conspicuous features Black bill and black feet / Less black on primary feathers than Red-billed Gulls / Shows more white on wings when in flight than Red-billed Gulls.

Call 'Kek kek kek kek' and 'kwaar kwaar kwaar' sounds, similar to Red-billed Gull.

Nest A colonial nester which makes a nest of twigs and seaweed on sand or river pebbles. Two pale-greenish, brown-blotched eggs are laid.

Target localities Miranda, Firth of Thames / Lake Taupo waterfront / Lake Rotorua near the Government Gardens, Rotorua / Most inland areas of Southland.

Red-billed Gull

Conspicuous colours Grey, white and black.
Habitat Coast.
Range Around New Zealand.
Size 370 mm (House Sparrow 145 mm, Fluttering Shearwater 300 mm).
Abundance Common.
Status Native.

Family Laridae Species *Larus novaehollandiae*

Common names Red-billed Gull, Tarapunga, Silver Gull (Australia)

Description – mature Head, neck and underparts White. Wings Grey, black-tipped on upper and undersides but with white markings on the black. Bill and legs Red. Eye Iris white, with a red ring around it.

Description – immature Wings Soft grey, with brownish speckles. Bill, legs and feet Black or brownish-black. Iris and eye-ring Brown.

Conspicuous feature Red bill and legs of adults.

Conspicuous characteristic Immature birds utter persistent food-begging, cheeting cries.

Call 'Kek kek kek' associated with strident screams and 'kwaar kwaar kwaar' sounds.

Nest A colonial nester, making a mound of sticks and seaweed on a sand bank or rock ledge. Up to two light-brown, dark brown-blotched eggs are laid. Sometimes they nest alongside Black-billed Gulls and White-fronted Terns.

Target localities Most beaches have Red-billed Gulls.

Differences between mature Red-billed Gulls and Black-billed Gulls

- Red-billed has a red bill; Black-billed has black bill.
- Red-billed has a shorter and heavier bill than Black-billed.
- Red-billed has red legs; Black-billed has reddish-black legs.
- Red-billed has distinctive black wingtips with a triangle of black on the end.
- Red-billed has a heavier head with the eye appearing to be less well balanced.
- Black-billed has only an edge of black on the wingtips, showing much more white on the wings when in flight.
- Black-billed has a softer appearance.
- Some Black-billed Gulls have black tail dots on the end of the upper tail.

Differences between immature Red-billed and Black-billed Gulls

- Red-billed has a black bill in the fledgling stage and blackish legs.
- Black-billed has a reddish bill with a dark tip and dark reddish legs in the fledgling stage.

Ecological differences between Red-billed and Black-billed Gulls

- Red-billed Gull takes a wide range of sea foods. It is also a recognised scavenger known for its rubbish-tip feeding habits.
- Black-billed Gulls are specialised feeders, taking estuarine fish and crustacean matter as well as worms and insects common on pasture. It has a regular habit of following the plough in parts of the South Island.

Habitat and nesting differences of Red-billed and Black-billed Gulls

- Red-billed Gulls are mainly coastal birds; Black-billed Gulls are predominantly inland birds.
- Red-billed Gull is a coastal nesting species; Black-billed Gull nests inland on the wide riverbeds of the South Island. In the North Island it is recorded nesting on sand and shell-spits but also inland at Lake Rotorua.

White-fronted Tern

Conspicuous colours Grey, white and black.
Habitat Coastal.
Range Around New Zealand.
Size 400 mm (House Sparrow 145 mm, Red-billed Gull 370 mm).
Abundance Common.
Status Native.

Family Laridae Species *Sterna striata*

Common names White-fronted Tern, Tara, Kahawai Bird

Description – breeding Forehead White above bill. This distinguishes it from the Black-fronted Tern of the South Island and the Antarctic Tern of Snares Islands. Crown Black. Upperparts Soft grey, sometimes with a pink tinge to breast. Underparts White. Bill Black. Legs Dull red.

Description – non-breeding Crown Receded black cap at back of a white head.

Description – immature bird Crown Receded black cap at back of white head. Wings Light grey with dark grey markings.

Conspicuous features White forehead Forked swallow-like tail.

Call A high-pitched 'kee-eet kee-eet' or a soft 'zeet zeet' sound. It has other, more rasping notes used at times of communal feeding.

Nest A colonial nester, making a scrape in the sand or shell or using a rock crevice. Up to two green, lightly spotted brown eggs are laid.

Migration Juvenile White-fronted Terns migrate across the Tasman and winter around the southeast Australian coast and the northern coast of Tasmania.

Other terns The Arctic Tern (*S. paradisaea*) and Common Tern (*S. hirundo*) from the northern hemisphere, and the Antarctic Tern (*S. vittata*), which breeds on the Snares Islands, can occasionally be seen in northern waters.

Target localities Many coastal beaches have this species often seen with gulls Miranda, Firth of Thames.

Differences between Black-fronted Terns and White-fronted Terns

- Black-fronted Tern has an orange bill and feet; White-fronted has a black bill and black legs.
- Black-fronted Tern has bluish-grey underparts; White-fronted has clean white underparts.
- Black-fronted Tern has black coming right to the bill; White-fronted has a white forehead above the bill.
- Black-fronted Tern is generally an inland bird except during winter months; White-fronted is coastal.

White-fronted Tern

Black-fronted Tern

Caspian Tern

Conspicuous colours Grey, white and black.
Habitat Coastal but sometimes seen over inland lakes.
Range Around New Zealand.
Size 510 mm (House Sparrow 145 mm, Red-billed Gull 370 mm).
Abundance Common.
Status Native.

Family Laridae

Species *Sterna caspia*

Common names
Caspian Tern, Tara-nui

Description — breeding
Crown Black to below the eye.
Upperparts Grey.
Underparts White.
Bill Bright red.
Feet and legs Black.

Description — non-breeding
Crown White, flecked with black through eye and round back of head.

Conspicuous features
Larger than the White-fronted Tern with which it frequently associates. The black cap, extending from the bill base on forehead, separates it from the smaller White-fronted Tern. Heavy bill. Body looks long and legs short.

Conspicuous characteristics Has a squat posture when standing. Usually feeds alone, flying slowly above water with head down. The heavy bill is obvious in this position. When a fish is spotted, birds hover momentarily and then dive forcefully in a similar fashion to gannets. Birds rest on sandbanks and high-tide roosts, often among shorebirds such as oystercatchers. When resting they spread through them.

Call A guttural 'kaar-kaa kaar-kaa kaar' noise, different from the higher pitched 'kee-eet' call of the White-fronted Tern. Immature birds make a whistle.

Nest A colonial nester which makes a scrape in the sand. Two greyish, brown-blotched eggs are laid.

Target localities Miranda, Firth of Thames. Seen on many beaches.

Grey-backed Storm Petrel

Conspicuous colours Grey, white and black.
Habitat Sea.
Range Waters around New Zealand from the south to Taranaki and Gisborne.
Size 180 mm (House Sparrow 145 mm, Fluttering Shearwater 300 mm).
Abundance Common in southern waters.
Status Circumpolar.

Family Hydrobatidae Species *Oceanites nereis*

Common names Grey-backed Storm Petrel, Reoreo

Description Head, neck, throat and upper breast Greyish-black. Upperwings Dark grey with black primary feathers. Rump and tail Dark grey with black tail tip. Underparts White. Legs and feet Dark grey.

Conspicuous features Feet extend behind tail when in flight / Tail is square / Underparts clean white.

Conspicuous characteristics Touches the sea with its feet when feeding / Dips constantly into the sea / Plunges erratically towards the water and then into the air again.

Call Cackle and twitter sounds near the breeding colonies.

Major breeding islands near New Zealand On islets off Chatham Islands, Antipodes Islands, Auckland Islands and Campbell Island.

Breeding months August to April. Nest is a shallow cup of grass and twigs above ground. One white egg is laid.

Range worldwide From New Zealand around southern hemisphere waters.

Target localities Waters off Kaikoura / Waters off Stewart Island.

White-faced Storm Petrel

Conspicuous colours Grey, white and black.
Habitat Sea.
Range Waters around New Zealand.
Size 200 mm (House Sparrow 145 mm, Fluttering Shearwater 300 mm).
Abundance Common.
Status Native.

Family Hydrobatidae Species *Pelagodroma marina*

Common names White-faced Storm Petrel, New Zealand White-faced Storm Petrel, Takahikare-moana

Description Face White, with a grey bar through and under eye. Upperparts Greyish-brown. Upperwings Dark grey with black primaries. Rump and tail Pale grey with black tail. Underparts White. Legs Black with yellow webs to feet.

Conspicuous features White on face · Yellow webs of feet · Slightly forked tail.

Conspicuous characteristics Appears to walk on the water with feet touching the waves · Skips and jumps across waves.

Call Cackle and twitter sounds near the breeding colonies.

Major breeding islands near New Zealand Offshore islands from Three Kings in the north to Stewart Island and Chatham Islands.

Breeding months August to April. One white egg is laid in a burrow.

Range worldwide From New Zealand around southern hemisphere waters.

Target localities Waters off Sandspit to Little Barrier Island · Waters off Kaikoura · Waters off Stewart Island.

Fairy Prion

Conspicuous colours Grey, white and black.
Habitat Sea.
Range Waters around New Zealand.
Size 250 mm (House Sparrow 145 mm, Fluttering Shearwater 300 mm).
Abundance Common.
Status Native.

Family Procellariidae

Species *Pachyptila turtur*

Common names
Fairy Prion, Titi wainui

Description
Face Grey with a faint whitish stripe above the eye.
Upperparts Bluish-grey.
Wings Bluish-grey with a black, open M marking across wings and lower back.
Underwing White.
Underparts White.
Tail Grey, with half the tail being boldly tipped with black.
Bill Black.

Conspicuous features
Dark, open M marking on upperwings.

Conspicuous characteristics A fast-flying species ✎ Birds toss and swerve above the wave-tops ✎ Birds alight on the water to feed.

Major breeding islands near New Zealand Offshore islands from Three Kings in the north to Stewart Island and Chatham Islands.

Breeding months November to February. One white egg is laid in a burrow.

Range worldwide From New Zealand around southern hemisphere waters.

Target localities Waters off Sandspit to Little Barrier Island ✎ Waters off Kaikoura ✎ Waters off Stewart Island.

Differences between Fairy Prion and Broad-billed Prion

- Fairy Prion has a narrow bill; Broad-billed Prion has a wide bill.
- Fairy Prion has a white underwing; Broad-billed has dark grey edges to primary feathers.
- Fairy Prion has pale eye surrounds; Broad-billed has a dark face and crown.
- Fairy Prion has a bold black tail tip; Broad-billed has a narrow black tip.

Note on prions

All six recognised species of prion are known from New Zealand waters. These are:

- Fairy Prion (*Pachyptila turtur*), a New Zealand species breeding on offshore islands from the Poor Knights in the north to Stewart Island.
- Fulmar Prion (*P. crassirostris*), which breeds on the Snares Islands and other southern islands.
- Thin-billed Prion (*P. belcheri*), which breeds from the Macquarie Islands (small colony only) and east to the Indian Ocean.
- Antarctic Prion (*P. desolata*), which breeds on southern islands from the Auckland Islands eastwards. This species is a common winter visitor to the northern seas about New Zealand.
- Salvin's Prion (*P. salvini*), which breeds on Indian Ocean islands but straggles to New Zealand.
- Broad-billed Prion (*P. vittata*), which breeds in abundance around the Chatham Islands and Stewart Island.

Similarities

All species of prion are difficult to identify in normal sea conditions. They are all of blue-grey colouring with white underparts and black tips to their tails. All have a black, open M mark across the wings and back.

Thin-billed Prion

Conspicuous colours Grey, white and black.
Habitat Sea.
Range Around New Zealand.
Size 260 mm (House Sparrow 145 mm, Fluttering Shearwater 300 mm).
Abundance Common.
Status Migrant.

Family Procellariidae

Species *Pachyptila belcheri*

Common name Thin-billed Prion

Description Face Above eye white. Behind eye black. Upperparts Light bluish-grey. Underparts White, except for black tip of tail. Wings Grey with a black M marking across upperwings when in flight. Underwing White. Bill Light grey and thin.

Conspicuous features Open M wing marking ✎ Light-grey forehead ✎ White stripe above the eye ✎ Thin bill.

Conspicuous characteristics Birds in fast flight reveal their white underwing ✎ Fast, erratic flyers, swerving from side to side.

Breeding islands Subantarctic – South America and Indian Ocean.

Target localities Waters off Kaikoura ✎ Waters off Stewart Island.

Broad-billed Prion

Conspicuous colours Grey, white and black.
Habitat Sea.
Range Around New Zealand.
Size 280 mm (House Sparrow 145 mm, Fluttering Shearwater 300 mm).
Abundance Common.
Status Native.

Family Procellariidae Species *Pachyptila vittata*

Common names Broad-billed Prion, Parara

Description Face Above eye white. Under eye to behind eye black. Upperparts Bluish-grey. Underparts White, except for tip of tail, which is black. Wings Grey with a black, open M marking across wings when in flight. Underwing White with grey trailing edges and wingtips. Bill Dark grey, tapering to a broad base at the gape.

Conspicuous features Open M wing marking · Prominent high forehead · White stripe above dark eye · Bill tapers to a broad base · White underwing.

Conspicuous characteristic Fast and erratic flyers.

Major breeding islands near New Zealand Breeds on offshore islands about Fiordland, Stewart Island and Chatham Islands.

Breeding months August to January. One white egg is laid in a burrow or crevice.

Range worldwide From New Zealand around southern hemisphere waters.

Target localities Waters off Sandspit to Little Barrier Island · Waters off Kaikoura · Waters off Stewart Island.

Pycroft's Petrel

Conspicuous colours Grey, white and black.
Habitat Sea.
Range Waters off the east coast of Northland and the Bay of Plenty.
Size 280 mm (House Sparrow 145 mm, Fluttering Shearwater 300 mm).
Abundance Rare.
Status Endemic.

Family Procellariidae

Species *Pterodroma pycrofti*

Common name Pycroft's Petrel

Description Head Dark-grey crown with darkish forehead feather scalation. Face, throat and around bill White with dark grey through the eye. Upperparts Grey. Upperwings Grey, with black, open M marking across wings and rump to wingtips. Underparts White. Underwings White, with black edges and a small, black tag at the leading wing-joint. Tail Grey with black tip. Legs and feet Bluish.

Conspicuous characteristics Graceful flier ✎ Flies with stiffly held wings ✎ Glides in wide circles.

Call 'Kek kek kek kek' made when near the breeding island.

Major breeding islands near New Zealand Poor Knights Islands, Hen and Chicken Islands and Mercury Islands.

Breeding months November to April. One white egg is laid in a rat hole-like burrow.

Range worldwide Migrates to central Pacific.

Target localities Waters from Sandspit to Little Barrier Island ✎ Waters off Coromandel Peninsula.

Differences between Pycroft's and Cook's Petrel (from Hadoram Shirihai)

- Pycroft's is slightly smaller than Cook's and its plumage overall appears a darker dusky grey.
- Pycroft's has almost no white supercilium stripe above the eye, and its head and nape appear darker and its neck shows more grey. Pycroft's has less abdomenal white showing when sitting on the water.
- In flight, Pyrcroft's looks shorter with more rounded wings and a longer tail.

Cook's Petrel

Conspicuous colours Grey, white and black.
Habitat Sea.
Range Waters off the east coast of North Island and around Stewart Island and Southland.
Size 290 mm (House Sparrow 145 mm, Fluttering Shearwater 300 mm).
Abundance Uncommon.
Status Endemic.

Family Procellariidae

Species *Pterodroma cookii*

Common names Cook's Petrel, Titi

Description Forehead White. Head Dark grey, with feather scalation above forehead. Face, throat and around bill White, with a dark grey stripe through eye. Upperparts Light grey. Upperwings A black M marking extends across wings and rump. Underparts White. Underwings White, with dark-grey edges and a dark tag at the leading edge wing-joint. Tail Light grey with dark grey tip. Bill Black. Legs and feet Bluish.

Conspicuous features Grey scalation on crown ✎ Black, open M wing marking.

Conspicuous characteristics Birds sit in small rafts on the sea ✎ In flight, birds tip and toss and show their white underparts and their wing M markings.

Call A distinctive 'ti ti ti ti', often heard after dark over Northland from birds returning to Little Barrier Island after feeding in the Tasman Sea.

Major breeding islands near New Zealand Great Barrier Island, Little Barrier Island, Codfish Island off Stewart Island.

Breeding months October to May. One white egg is laid in a deep burrow.

Range worldwide Migrates to northeast Pacific.

Target localities Waters from Sandspit to Little Barrier Island ✎ Waters off Coromandel Peninsula ✎ Waters off Stewart Island.

Black-winged Petrel

Conspicuous colours Grey, white and black.
Habitat Sea.
Range Waters around New Zealand but not Southland.
Size 300 mm (House Sparrow 145 mm, Fluttering Shearwater 300 mm).
Abundance Common.
Status Native.

Family Procellariidae

Species *Pterodroma nigripennis*

Common name Black-winged Petrel

Description Forehead White, with pale-grey feather scalation to pale-grey crown. Head and back Pale grey, the grey forming an incomplete collar on the neck. Eye Dark patch to behind eye. Upperwings Grey with black, open M marking across wings and rump from wingtip to wingtip. Underparts White. Underwings White with black edges, and tabs at the wing-joint. Tail Grey with dark-grey tip. Undertail White with dark-grey tip. Bill Black. Feet Pink.

Conspicuous features Dark underwing leading edges.

Conspicuous characteristics Will fly over breeding colonies in daylight. Attracted to shipping and will land on decks.

Call An 'ahhoo – wiwiwiwi' sound is heard at dusk.

Major breeding islands near New Zealand Kermadec Islands, Three Kings, Motuopao Island off Cape Maria Van Diemen, South East Island near Chatham Islands.

Breeding months December to June. One white egg is laid in a deep burrow.

Range worldwide Migrates to northeast Pacific.

Target localities Waters off Cape Reinga. Waters from Sandspit to Little Barrier Island.

Mottled Petrel

Conspicuous colours Grey, white and black.
Habitat Sea.
Range Waters to the south of the South Island.
Size 340 mm (House Sparrow 145 mm, Fluttering Shearwater 300 mm).
Abundance Common.
Status Endemic.

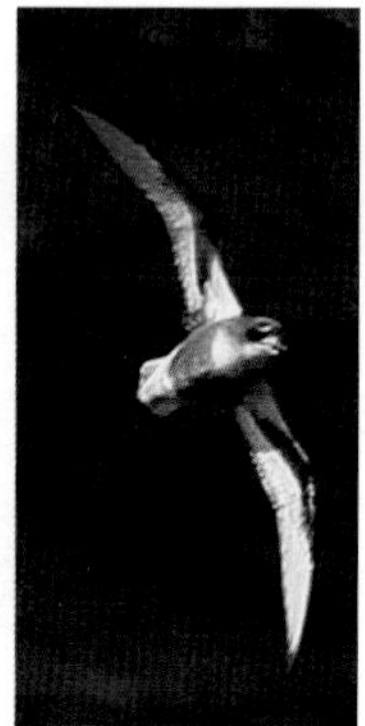

Family Procellariidae **Species** *Pterodroma inexpectata*

Common names Mottled Petrel, Korare

Description Crown Grey with dark-grey feather scalation above bill. Upperparts and back Dark grey with white horizontal mottling. Upperwings Dark grey with black, open M marking from wingtip, across rump, to wingtip. Uppertail Mottled grey. Underparts White with prominent grey abdomen patch. Underwings White with black edges and bold black tab at the wing-joint. Tail Grey with light-black tip.

Call A high-pitched 'ti ti ti ti ti' uttered by returning birds at night.

Major breeding islands near New Zealand Codfish Island and islands around Stewart Island and Snares Islands.

Breeding months December to June. One white egg is laid in a burrow or crevice.

Range worldwide Migrates to northeast Pacific.

Target localities Waters around Stewart Island.

Differences between Mottled Petrel and Cook's Petrel

- Mottled Petrel is bigger, and has a dark grey abdominal patch and bolder underwing markings.
- Mottled Petrel has an overall darker upperwing and a darker M marking on the upperwing.

Antarctic Fulmar

Conspicuous colours Grey and white.
Habitat Sea.
New Zealand range Widely spread north to the Bay of Plenty on the east coast and to the Kaipara Harbour on the west. At Kaikoura can be seen feeding close to land especially when the weather is rough.
Size 500 mm (House Sparrow 145 mm, Flesh-footed Shearwater 450 mm).
Abundance Uncommon.
Status Native.

Family Procellariidae Species *Fulmarus glacialoides*

Common names Antarctic Fulmar, Silver-grey Fulmar

Description Head, neck and underparts White. Underwing White with black trim to trailing edges. Upperwings and mantle Soft grey, with dark edges to primary feathers. Tail White. Bill Pinkish with dark tip. Legs and feet Pinkish blue.

Conspicuous features Larger than Flesh-footed Shearwater / White head and tail.

Conspicuous characteristics Usually seen alone / Will join in seabird feeding frenzies.

Breeding islands close to New Zealand On islands at the fringes of the Antarctic.

Breeding months December onwards in colonies. One white egg is laid in a shallow scrape.

Range worldwide Around southern oceans.

Target localities Kaikoura coastline / Seas around Stewart Island.

Buller's Mollymawk

Conspicuous colours Grey, white and black.
Habitat Sea.
Range Around New Zealand.
Size 800 mm (House Sparrow 145 mm, Red-billed Gull 370 mm).
Abundance Common.
Status Endemic.

Family Diomedeidae Species *Diomedea bulleri*

Common name Buller's Mollymawk

Subspecies Two are recognised: Northern Buller's Mollymawk (*D. b. platei*) and Southern Buller's Mollymawk (*D. b. bulleri*).

Description – mature Head Grey, with a silvery grey cap on Northern Buller's, and a silvery white cap on Southern Buller's. Has a dark patch through upper eye. Upperwings and back Black. Rump White. Underwings White, with a cleanly defined, thickish black leading edge and a thin, black trailing edge. Underparts White. Tail White with broad, dark-grey tip. Bill Black-sided with bright yellow edges to top and bottom. The bill of Northern Buller's Mollymawk is slightly heavier.

Conspicuous features A handsome bird when sitting on the water — Smaller than Grey-headed and Shy Mollymawks — Bill similar to Grey-headed with bright yellow edges and black sides.

Major breeding islands near New Zealand Northern Buller's Mollymawk: Rosemary Rocks, Three Kings Islands (small population) and Motuhara Islands east of Chatham Islands. Southern Buller's Mollymawk: Snares Islands and Solander Island.

Breeding months Northern Buller's Mollymawk: October to June. Southern Buller's Mollymawk: January to October.

Range worldwide Around New Zealand and to South America.

Target locality Waters off Kaikoura — Waters off Stewart Island.

Grey-headed Mollymawk

Conspicuous colours Grey, white and black.
Habitat Sea.
Size 800mm (House Sparrow 145 mm, Red-billed Gull 370 mm).
Abundance Common.
Status Circumpolar.

Family Diomedeidae Species *Diomedea chrysostoma*

Common name Grey-headed Mollymawk

Description – mature bird Head Grey. Upperwings and back Black. Rump White. Underparts White. Tail Grey and black-tipped. Bill Black, edged with bright yellow above and below, and tipped with pink.

Conspicuous features Dark grey head similar to Buller's Mollymawk but lacks its white crown ✎ Shy Mollymawks have a range of slightly grey-headed birds but they are bigger ✎ Bill is black, edged with bright yellow above and below, and tipped with pink ✎ Underwing shows more white than Black-browed but less than Buller's ✎ The leading underwing edge shows a wider belt of black than the trailing black edge.

Major breeding islands near New Zealand Campbell Island.

Breeding months September to May.

Range worldwide Throughout the southern oceans.

Target locality Waters from Sandspit to Little Barrier Island ✎ Waters off Kaikoura.

Salvin's Mollymawk

Conspicuous colours Grey, white and black.
Habitat Sea.
Range Around New Zealand and the southern oceans.
Size 900mm (House Sparrow 145 mm, Red-billed Gull 370 mm).
Abundance Common in northern waters.
Status Endemic.

Family Diomedeidae Species *Diomedea cauta salvini*

Common names Salvin's Mollymawk, Bounty Island Mollymawk

Description – mature Head Dark grey with black through eye and whitish cap. Upper wings Black. Back Grey-black merging into darker grey. Rump and underparts White. Underwings White with thin black edging and black primary tips. Tail White, tipped with a broad band of dark grey. Bill Dark grey-sided, with yellow edges top and bottom but the yellow on the bottom mandible giving way to a black patch at the tip.

Conspicuous features Grey head and neck · Whitish cap · Bill differs from the Shy Mollymawk, being dark grey-sided with a line of yellow under bottom mandible.

Major breeding islands near New Zealand Bounty Islands and Snares Islands.

Breeding months September to August.

Range worldwide Around New Zealand to South Africa and the Indian Ocean.

Target localities Cook Strait · Waters off Stewart Island.

Chatham Island Mollymawk

Conspicuous colours Grey, white and black.
Habitat Sea.
Range Around Chatham Islands.
Size 900 mm (House Sparrow 145 mm, Shy Mollymawk 900 mm).
Abundance Uncommon.
Status Endemic.

Family Diomedeidae Species *Diomedea cauta eremita*

Common name Chatham Island Mollymawk

Description — mature Head and neck Dark grey with dark patch through eye. Upperwings and back Black. Rump White. Underparts White. Underwings White, thinly edged with black and with dark primary tips. Tail White, tipped with a broad band of dark grey. Bill Bright yellow, with black patch at end of lower mandible.

Conspicuous features Head and neck are all-over dark grey Bill bright yellow with black tip to lower mandible. This separates it from both the Shy and Salvin's.

Breeding islands close to New Zealand The Pyramid at Chatham Islands.

Breeding months September to April.

Range worldwide Waters of the Chatham Islands group.

Target localities The Pyramid and waters around the Chatham Islands Occasionally strays to the South Island.

Galah

Conspicuous colour Pink.
Habitat Open country, trees.
Range An Australian cage-bird escapee that is establishing itself in the South Auckland areas of Clevedon, Kawakawa Bay, Mangatangi, Mangatawhiri and Waitakaruru. Also on Ponui Island off Kawakawa Bay near Auckland.
Size 360 mm (House Sparrow 145 mm, Kaka 450 mm).
Abundance Rare.
Status Introduced.

Family Cacatuidae **Species** *Cacatua roseicapilla* **Common name** Galah

Description Crown White. Upperwings Grey. Neck, abdomen and underwing coverts Pink. Underwing Grey with dark primary feathers. Rump White. Undertail Grey. Bill Off-white.

Conspicuous features Pink abdomen extending to underwing coverts and neck ✎ White crown.

Conspicuous characteristics Often in fast-moving flocks outside of breeding season ✎ Noisy when in flocks ✎ Feeds among maize stubble.

Call A high-pitched 'chi chi chi'.

Nest Nests in a hole in an old tree. Up to five white eggs are laid.

Target localities Ponui Island and Pakihi Island in Auckland Harbour ✎ Clevedon, South Auckland ✎ Mangatawhiri, South Auckland.

Red-necked Stint (breeding)

Conspicuous colours Brown and white.
Habitat Coast and estuaries.
Range Scattered throughout New Zealand.
Size 150 mm (House Sparrow 145 mm, Wrybill 200 mm).
Abundance Rare.
Status Migrant.

Family Scolopacidae **Species** *Calidris ruficollis*
Common name Red-necked Stint

Description – breeding Face, neck, chin and throat Brick red. New Zealand birds show only slight reddening prior to departure. Upperparts Dark brown with rufous edges to feathers. Underparts White.

Description – non-breeding Forehead and eyebrow White. Head and upperparts Grey with brownish tinge. Neck Traces of red on sides. Wings Greyish, with a white line across the middle of the wings. Tail Whitish outer feathers with black inner. Underparts White. Bill and legs Black.

Conspicuous features The smallest of the Arctic-breeding wading birds to come to New Zealand ✎ Bill is black, short and straight.

Conspicuous characteristics Prefers to feed on areas of mud covered by a thin film of water ✎ Feeds busily with sewing-machine action ✎ Runs here and there ✎ Usually found towards the edges of Wrybill flocks ✎ Often feeds in the company of other sandpipers and Wrybills ✎ In flight, birds fly low and direct in a compact group.

Target localities Miranda, Firth of Thames ✎ Maketu Harbour, Bay of Plenty ✎ Embankment Road, Lake Ellesmere, near Christchurch ✎ Waituna Lagoon near Invercargill.

Curlew Sandpiper (breeding)

Conspicuous colours Red and brown
Habitat Coast and estuaries.
Range Scattered throughout New Zealand.
Size 190 mm (House Sparrow 145 mm, Wrybill 200 mm).
Abundance Rare.
Status Migrant.

Family Scolopacidae **Species** *Calidris ferruginea*

Common name Curlew Sandpiper

Description – breeding (note that female birds are paler) Face White around base of bill and above eye, otherwise brick-red. Head and neck Brick-red. Back and wings Red with flecks of black and silver. Breast and underparts Brick-red with black barring. Rump, undertail and underwing White.

Description – non-breeding Face Greyish-brown with white eyebrow. Upperparts Grey-brown. Wings Brown, with a thin white line across middle of wings. Rump White. Breast Greyish wash. Underparts and underwing White. Tail White with brown tip. Bill Black, noticeably down-curved at the tip. Legs Black.

Conspicuous features Brick-red colouring over much of the body ✎ Down-curved tip to the bill ✎ White rump of bird in flight ✎ White wing line on upperwings of bird in flight.

Conspicuous characteristics Usually feeds among Sharp-tailed Sandpipers or Wrybills ✎ Tends to start feeding immediately the tide has turned ✎ Very fast and manoeuvrable flier when in flocks.

Call A gentle chirrup sound.

Target locality Miranda, Firth of Thames ✎ Manawatu Estuary, Foxton.

Banded Dotterel (breeding)

Conspicuous colours Red and brown.
Habitat Coast, estuaries and inland riverbeds.
Range Throughout New Zealand except for King Country and Fiordland.
Size 200 mm (House Sparrow 145 mm, Wrybill 200 mm).
Abundance Common.
Status Endemic.

Family Charadriidae **Species** *Charadrius bicinctus*

Common names Banded Dotterel, Tuturiwhatu, Double-banded Plover (Australia)

Description – male (breeding) Forehead White patch above bill with black patch above white. Head Brown with black line from bill to under eye. Upperparts Brown. Chin White. Underchin band Black above, white below. Breast band Rich chestnut. Underparts White.

Description – female (breeding) Overall appearance Similar to male but paler with the lower breast band being narrower and less distinct.

Description – immature Overall appearance Paler than adult birds, no breast band and of fawny appearance. Shoulder-tabs Brown.

Description – wintering Overall appearance Both sexes are dull, retaining a partial upper breast band. By July full breeding plumage becomes apparent.

Conspicuous features Smaller than the New Zealand Dotterel ✎ Double bands on breeding plumaged birds.

Conspicuous characteristics Runs and pauses when feeding ✎ Often seen in large flocks standing motionless in ploughed paddocks.

Call A loud 'pit pit' and a trilling 'che-ree-a-ree'.

Nest A scrape on the sand. Up to three greenish, black-spotted eggs are laid.

Target localities Kawakawa Bay, east of Papakura ✎ Miranda, Firth of Thames ✎ Ahuriri River near Twizel ✎ Lake Gunn, Eglinton Valley.

Lesser Knot (breeding)

Conspicuous colours Red and brown.
Habitat Harbours and estuaries.
Range Throughout New Zealand.
Size 250 mm (House Sparrow 145 mm, Wrybill 200 mm).
Abundance Common.
Status Migrant.

Family Scolopacidae **Species** *Calidris canutus*

Common names Lesser Knot, Red Knot, Huahou

Description – breeding plumage Head, neck and breast Chestnut. Upperparts A mixture of black, dark brown, grey and white. Abdomen Rufous.

Description – non-breeding plumage Upperparts and wings Grey with thin, white wingbar. Underparts Greyish-white with some speckling on flanks. Rump Barred with white and grey. Bill Brown. Legs Greyish-brown.

Conspicuous features Rufous abdomen ✎ Chestnut underparts ✎ White wingbar when in flight ✎ Squat posture.

Conspicuous characteristics Intermingles with godwits on high-tide roosts often in the middle of the mob.

Call Usually silent. Before migration often a two syllable 'poor-me poor-me' sound is made.

Target localities Clarks Beach, Manukau Harbour, out from Papakura ✎ Miranda, Firth of Thames ✎ Motueka Estuary, Marlborough ✎ Awarua Bay, Invercargill.

New Zealand Dotterel (breeding)

Conspicuous colours Red and brown.
Habitat Coast and estuaries.
Range North of the North Island and on Stewart Island.
Size 250 mm (House Sparrow 145 mm, Wrybill 200 mm).
Abundance Uncommon.
Status Endemic.

Family Charadriidae

Species
Charadrius obscurus

Common names
New Zealand Dotterel,
Tuturiwhatu,
Red-breasted Dotterel

Subspecies Two are recognised: Northern New Zealand Dotterel (*C. o. aquilonius*) and Southern New Zealand Dotterel (*C. o. obscurus*).

Description – breeding
Breast Male birds attain red breasts and abdomens in July. Female birds attain lesser amounts of red colouring.

Description – non-breeding Head Brown, with a white forehead and white eye-stripe. Upperparts Brown. Underparts White. Bill Black and robust. Legs Grey.

Conspicuous features Red breast in breeding season ✎ White eye-stripe ✎ Dark eye and bill.

Conspicuous characteristics For much of the year it seldom leaves the breeding territory ✎ Blends in well with the high-tide roost and breeding territories.

Call A 'kreek kreek' call when in flight.

Nest A scrape in the sand with minimal nest lining. Up to three buff-brown, dark-brown blotched eggs are laid. In the North Island nests are usually found behind the beach in locations which get 360° views. Very often mated pairs occupy territories at the opposite ends of beaches. On Stewart Island birds nest high on the island in prominent positions with good visibility, although there are records of sand-dune nests from Mason Bay.

Target localities Kawakawa Bay, east of Papakura ✎ Miranda, Firth of Thames ✎ Opoutere Spit, Coromandel ✎ Maketu Harbour, Bay of Plenty ✎ Riverton Beach, Southland.

Bar-tailed Godwit (breeding)

Conspicuous colours Red and brown.
Habitat Coast and estuaries.
Range Throughout New Zealand.
Size 390 mm (House Sparrow 145 mm, Wrybill 200 mm).
Abundance Common.
Status Migrant.

Family Scolopacidae

Species *Limosa lapponica*

Common names Bar-tailed Godwit, Eastern Bar-tailed Godwit, Kuaka

Description – breeding (male) Head Brick-red with whitish stripe above the eye. Upperparts Dark brown with buff edges to feathers. Neck and underparts Chestnut. Underwing White. Tail Brown barred with white. Bill Brownish-pink.

Description – breeding (female) Upperparts Greyish streaked with brown. Breast Buff lightly tinged with red. Underparts White with fine barring on edges of abdomen.

Description – non-breeding Upperparts Brown streaked with dark brown. Underparts White. Rump and tail White barred with brown. Bill Brownish-pink with a black tip. Legs Grey.

Conspicuous features A large wading bird ✎ Females are larger than male birds ✎ Bill is long and up-curved ✎ Females have longer bills than males.

Conspicuous characteristics Flies in loose and straggly skeins ✎ Congregates in large numbers on high-tide roosts ✎ Feeds busily once the tide recedes ✎ Digs bill deeply into mud to feed, so often seen with a muddy face ✎ Males develop rufous tonings towards the end of summer ✎ Auckland birds are often seen changing harbours in straggly skeins.

Godwits and Knots

Call A soft 'kit kit kit' or a 'kew kew' sound. Just prior to migration birds become excited and noisy at high-tide roosts.

Arrival dates From mid-September.

Departure dates Birds congregate in northern harbours from February onwards at places like the Manukau Harbour and the Firth of Thames. They start leaving from mid-March.

Overwintering birds At any time of year including winter, godwits in small numbers, often in breeding plumage, can be seen in many harbours.

Target localities Auckland muddy beaches ✎ Kawakawa Bay, east of Papakura ✎ Miranda, Firth of Thames ✎ Motueka Estuary, Nelson ✎ Fortrose Estuary, east of Invercargill.

Godwits and Gulls

Black-tailed Godwit (breeding)

Conspicuous colours Red and brown.
Habitat Coast and estuaries.
Range Scattered throughout New Zealand.
Size 390 mm (House Sparrow 145 mm, Wrybill 200 mm).
Abundance Rare.
Status Migrant.

Family Scolopacidae **Species** *Limosa limosa*

Common names Black-tailed Godwit, Eastern Bar-tailed Godwit, Kuaka

Description – breeding Head, neck and breast Chestnut. Eyebrow Whitish. Upperparts Brownish with individual feathers having buff edges. Abdomen White with some side barring with brown. Legs Lead-grey.

Description – non-breeding Upperparts Smooth grey. Upperwings Grey, with black edges to the secondary feathers, under which is a white band which passes through the black primaries. Underwing White edged with black. Underparts Whitish. Rump White, the white extending down the tail to a broad, black tip. Bill Greyish with a black tip.

Conspicuous features Similar in size to a Bar-tailed Godwit White upperwing marking White tail and black tail-tip band Straight bill, not up-curved like the Bar-tailed Godwit.

Conspicuous characteristics Legs trail behind the bird in flight Legs longer than those of the Bar-tailed Godwit Is easily scared and quickly takes to the air when disturbed.

Call When put to flight a loud 'wicka wicka wicka' call is made.

Target locality Miranda, Firth of Thames.

Goldfinch

Conspicuous colours Red, yellow, blue and black.
Habitat Parks, gardens, forests.
Range Throughout New Zealand.
Size 130 mm (House Sparrow 145 mm).
Abundance Common.
Status Introduced.

Family Fringillidae **Species** *Carduelis carduelis* **Common name** Goldfinch

Description Crown Centre crown and back of head black. Face Red with white behind the eye and on chin. Back Reddish-brown. Wings Primaries and secondaries black with a gold bar between and some white on ends of feathers. Breast White with a brown wash. Underparts White. Rump White. Tail Black, spotted white near the tip.

Conspicuous features Sexes are alike ✎ Red, white and black on face ✎ Gold bar on wings.

Conspicuous characteristics Forms into large flocks after the breeding season ✎ Regularly feeds on the ground on sow thistle seed (*Sonchus oleraceus*) and grass seed (*Poa annua*), but also high up in trees such as alder trees (*Alnus glutinosa*) ✎ Flies with undulating flight, showing slight forked tail ✎ Utters a fragile twitter while in flight. Don't confuse with Redpoll flight-song which is loud and clear in comparison.

Call Song, delivered by the male bird, is a rather thin but persistent jingling twitter usually from power wires or a high perch. Song period is from October to February.

Nest A neatly made cup of fine twigs and grasses, camouflaged with silver lichen, and lined with feathers or wool, placed on a branch or in a fork of a tree, high up. Up to five blue, brown-spotted eggs are laid.

Target localities Parks and gardens ✎ Miranda, Firth of Thames ✎ Common everywhere.

Chaffinch (male)

Conspicuous colours Red, yellow, blue and black.
Habitat Parks, gardens, forests.
Range Throughout New Zealand.
Size 150 mm (House Sparrow 145 mm).
Abundance Common.
Status Introduced.

Family Fringillidae **Species** *Fringilla coelebs* **Common name** Chaffinch

Description – male Forehead Black. Crown and nape Bluish-grey. Face, throat and underparts Pinkish-brown. Back Reddish-brown. Rump Olive. Wings Black with two conspicuous white wingbars on each wing. Tail Central feathers brown. Outer feathers black with white outer edges.

Description – female Upperparts Brownish-grey. Wings As for male bird. Rump Green. Underparts Greyish-brown. Tail As for male bird.

Conspicuous features Bluish-grey crown ✎ Pinkish-brown around face, throat and upper breast ✎ Shows a lot of white on wings and tail when in flight.

Conspicuous characteristics Feeds on the ground but flies up for cover when disturbed ✎ In flight, when in forests, it is faster and more direct than native birds ✎ Has a habit of hawking insects like a flycatcher, especially over water. Don't confuse with a swallow ✎ Continually sings during the breeding season ✎ During winter and early spring joins mixed flocks of other finches.

Call Song delivered by the male bird is a happy 'cherry cherry cherry cheery swee swee' repeated many times throughout the day. Throughout the year both sexes utter a metallic 'pink pink' note. Song period is July to January.

Nest A neatly constructed cup of dry grass and moss, camouflaged with silver lichen, at about 3 m from the ground. Up to four bluish, purple-blotched eggs.

Target localities Parks and gardens.

Eastern Rosella

Conspicuous colours Red, yellow, blue and black.
Habitat Parks, gardens, forests.
Range Northern North Island and Wellington area. Also Otago.
Size 320 mm (House Sparrow 145 mm).
Abundance Common.
Status Introduced.

Family Psittacidae

Species *Platycercus eximius*

Common name Eastern Rosella

Description
Head, neck and breast Bright red.
Lower cheeks and throat White.
Wings A mixture of yellow and black with blue primaries.
Underparts Yellow.
Tail Blue and green.

Conspicuous feature Brightly coloured.

Conspicuous characteristics Usually seen flying in pairs. A difficult bird to see once it has landed in a tree, as it tends to move into the middle of the tree. Birds often call from high up.

Call The usual call is a metallic 'kwink' heard from high up in the forest. It also has an attractive musical song.

Nest In cavities in trees. Up to six white eggs are laid.

Target localities Wenderholm Regional Park, North Auckland. Cornwall Park, Auckland.

Crimson Rosella

Conspicuous colours Red, yellow, blue and black.
Habitat Parks, gardens, forests.
Range Wellington area.
Size 350 mm (House Sparrow 145 mm).
Abundance Rare.
Status Introduced.

Family Psittacidae

Species *Platycerus elegans*

Common name Crimson Rosella

Description — mature
Head and underparts Red.
Face and throat Blue.
Wings Black and red with blue primaries.
Tail Light-blue outer feathers and dark-blue inner feathers.

Description — immature
Head Crimson.
Face and throat Blue.
Upper and underparts Green.
Wings Green, black-blotched with blue primaries. Tail Crimson, with blue outer feathers.

Conspicuous features General appearance of a mature bird is of an all-over red bird ✎ Red and blue colouring.

Conspicuous characteristics Will associate with Eastern Rosellas ✎ Direct, fast-flying flight, with outstretched tail.

Call A loud 'kwink kwink kwink', similar but of deeper notes than the Eastern Rosella. It also has a musical song not unlike the Eastern Rosella.

Nest Cavities in trees about 5 m up. Up to six white eggs are laid.

Target localities Botanic Garden area, Wellington ✎ Karori, Wellington.

Whitehead

Conspicuous colours White and brown.
Habitat Forests.
Range North Island from Waikato south, and on some offshore islands.
Size 150 mm (House Sparrow 145 mm).
Abundance Common in selected areas.
Status Endemic.

Family Pachycephalidae

Species *Mohoua albicilla*

Common names Whitehead, Popokotea, Bush Canary

Description Head, neck and underparts White with male birds being whiter than female birds. Wings, back and uppertail Pale brown. Bill, legs and feet Black. Eye Black.

Conspicuous features White head and underparts ✎ Black eye.

Conspicuous characteristics Feeds from the middle of the forest to high in the canopy ✎ Has a habit of working up tree trunks, usually in a hurry ✎ Often sings in short, melodic bursts as it feeds ✎ Makes short, almost hopping-like flights, from treetop to treetop ✎ In winter it forms noisy flocks.

Call Has two songs, one a canary-like chatter made up of 'swee swee swee chir chir chir' notes, and the other a melodic 'swee swee, ee swee o'.

Nest Cup-shaped, made of twigs bound with cobwebs and lined with feathers or tree-fern hair, usually built towards the end of branches, about 4 m above the ground. Up to four white, brown-speckled eggs are laid. Whitehead nests are parasitised by the Long-tailed Cuckoo, which lays one creamy-white egg.

Target localities Tiritiri Matangi Island, near Whangaparaoa ✎ Pine forests near Rotorua ✎ Lake Rotopounamu near Tokaanu ✎ Pureora Forest near Benneydale ✎ Waitomo Caves.

Barbary Dove

Conspicuous colour White.
Habitat Garden, suburbs and open country.
Range Established in a few selected areas.
Size 280 mm (House Sparrow 145 mm).
Abundance Uncommon.
Status Introduced. Originates from Asia and closely related to the Ring-necked Doves of Africa (*S. decipens* and *S. capicola*) and to the Collared Dove of Europe and Asia, (*S. decaocta*).

Family Columbidae **Species** *Streptopelia roseogrisea*

Common name Barbary Dove

Description Upperparts and underparts Creamy buff with brown wing edges. Neck Black collar across the back. Tail Fawn with white tips to outer tail feathers.

Conspicuous feature Brown neck collar, which separates it from the Spotted Dove.

Conspicuous characteristics Sits on telephone wires and television aerials. Feeds on lawns and driveways.

Call A soft 'kru ku'. Don't confuse with Morepork or the tri-syllabic call of the Spotted Dove.

Nest A twiggy, flimsy structure built at about 4 m from the ground. Two white eggs are laid.

Target localities Kerikeri near the Stone Store. Orewa at the north end. Puketutu Island across the causeway from the Mangere sewage plant, Auckland. Havelock North, Hawke's Bay.

Sulphur-crested Cockatoo

Conspicuous colour White.
Habitat Native forest remnants, exotic plantations and especially eucalypt trees.
Range The main North Island populations are found in small groups in areas to the west from Awakino in the south to Muriwai Beach in the north. Other populations exist near Miranda, Hunterville and Waikanae. In the South Island known from around Christchurch and south of Dunedin.
Size 500 mm (House Sparrow 145 mm, Kaka 450 mm).
Abundance Uncommon.
Status Introduced.

Family Cacatuidae
Species *Cacatua galerita*
Common name Sulphur-crested Cockatoo
Description
Crest and behind eye Yellow.
Upperparts and underparts White.
Bill and feet Greyish-black.
Eye Brown with a blue eye-ring.
Conspicuous feature Sulphur-coloured crest.
Conspicuous characteristics In the breeding season regularly seen in pairs. Out of the breeding season found in large, mobile, noisy flocks.
Call Harsh and raucous screeches.
Nest High up in hole in tree. Two white eggs are laid.

Target localities Miranda district, Firth of Thames; The cemetery, Ngarara Road, Waikanae; Greendale south of Christchurch.

Cattle Egret

Conspicuous colour White.
Habitat Pasture where cattle are grazing and wet areas.
Range Throughout New Zealand in selected areas.
Size 510 mm (House Sparrow 145 mm, Red-billed Gull 370 mm).
Abundance Uncommon.
Status Migrant.

Family Ardeidae

Species *Bubulcus ibis*

Common name
Cattle Egret

Description – non-breeding
Head and breast White.
Upper and underparts White.
Bill Yellow.
Legs Dark grey.

Description – breeding
Head and breast Rusty-coloured.
Upperparts and abdomen White.
Wings White with rusty-coloured plumes.
Bill Orange.
Legs Dark grey.

Conspicuous features A small, white heron ✎ Trailing dark-grey legs when birds are in flight ✎ Rusty breeding plumage.

Conspicuous characteristics Birds are gregarious, usually being found in large flocks ✎ Birds are easily scared and cannot be approached easily.

Call In New Zealand silent, but near the nest in Australia the call is a 'rick-rack rick-rack', or a 'kraah' sound.

Target localities Kopurahi, Hauraki Plains ✎ Rangiriri, North Waikato ✎ Lake Ngaroto, South Waikato.

Little Egret

Conspicuous colour White.
Habitat Estuaries.
Range Scattered throughout New Zealand in selected areas.
Size 560 mm (House Sparrow 145 mm, Red-billed Gull 370 mm).
Abundance Uncommon.
Status Migrant.

Family Ardeidae **Species** *Egretta garzetta* **Common name** Little Egret

Description Upperparts and underparts White. Bill Black. Legs Black. Feet Yellow.

Conspicuous features A small, all-white heron • Black bill and legs • In breeding plumage it has two long, narrow plumes, which fall from the nape, and also plumes on back and breast.

Conspicuous characteristics Usually a solitary bird but will associate with White Herons • When feeding it is very active, dashing in pursuit of prey.

Call Harsh croaks.

Nest A tree nester. Does not breed in New Zealand.

Target localities Miranda, Firth of Thames • Matata Lagoons, Bay of Plenty • Waikanae Estuary, north of Wellington • Waituna Lagoon, east of Invercargill.

Intermediate Egret

Conspicuous colour White.
Habitat Estuaries and inland waterways.
Range Scattered throughout New Zealand in selected areas.
Size 640 mm (House Sparrow 145 mm, Red-billed Gull 370 mm).
Abundance Rare.
Status Migrant.

Family Ardeidae

Species
Egretta intermedia

Common names
Intermediate Egret, Plumed Egret

Description – non-breeding
Upperparts and underparts White.
Bill Yellow.
Legs and feet Black.

Description – breeding
Upper and underparts White with white plumes on back.
Bill and legs Reddish.
Facial skin Green.

Conspicuous features
A small, all-white heron ✎ In breeding plumage has distinctive white plumes on back and breast ✎ Black line from gape ends level with eye. With the White Heron it extends beyond the eye.

Conspicuous characteristics A solitary bird ✎ Stands motionless among rushes and reeds when feeding.

Call Harsh croaks at nest.

Nest In reeds, vegetation or trees. Does not breed in New Zealand.

Target localities Rangiriri, North Waikato ✎ Matata Lagoons, Bay of Plenty ✎ Waituna Lagoon, east of Invercargill.

Royal Spoonbill

Conspicuous colour White.
Habitat Estuaries.
Range Scattered throughout New Zealand in selected areas.
Size 770 mm (House Sparrow 145 mm, Red-billed Gull 370 mm).
Abundance Common in selected areas.
Status Native.

Family Threskiornithidae **Species** *Platalea regia*

Common names Royal Spoonbill, Kotuku-ngutupapa

Description Face Black skin to behind eye with a yellow patch above each eye and a red spot in centre of forehead. Upperparts and underparts White. Bill Black and spoon-shaped. Legs Black.

Conspicuous features A large, white, heron-like bird ✎ Black, spoon-shaped bill.

Conspicuous characteristics Flies with neck and legs outstretched ✎ Bill shape visible when in flight ✎ Legs extend beyond body when in flight ✎ Gregarious feeders ✎ Constantly on the move when feeding ✎ Has a comical waddling gait when feeding caused by a feeding action of swinging the bill from side to side through water and soft mud ✎ Perches on prominent treetops or poles.

Call Soft, guttural grunts near the nest but usually silent. Sometimes soft bill chattering is heard.

Nest A platform of twigs in tall trees. Up to four white eggs, lightly blotched brown, are laid.

Target localities Miranda, Firth of Thames ✎ Manawatu Estuary, near Foxton ✎ Waikanae Estuary, north of Wellington ✎ All Day Bay Lagoon, south of Oamaru.

Feral Goose

Conspicuous colour White.
Habitat Open pasture, wetlands.
Range Scattered throughout New Zealand in selected rural areas.
Size 800 mm (House Sparrow 145 mm, Mute Swan 1500 mm).
Abundance Common in selected areas.
Status Introduced.

Family Anatidae

Species *Anser anser*

Common name Feral Goose

Description – male
Upperparts and underparts White.
Bill legs and feet Orange.

Description – female
Upperparts and underparts White.
Wings, neck and thighs Grey.

Conspicuous features
Grey wings on female birds ✎ Orange bill and legs ✎ Straight neck.

Conspicuous characteristic Usually with heads down grazing.

Call Loud honking sounds.

Nest Bowl in tall grass. Up to six cream eggs are laid.

Target localities Open countryside.

White Heron

Conspicuous colour White.
Habitat Estuaries.
Range Scattered throughout New Zealand in selected areas.
Size 920 mm (House Sparrow 145 mm, Red-billed Gull 370 mm).
Abundance Uncommon.
Status Native.

Family Ardeidae

Species *Egretta alba*

Common names
White Heron,
Kotuku,
Great White Egret (Australia)

Description
Upper and underparts White.
Bill Black in breeding birds but yellow in non-breeding birds.
Legs and feet Black. Yellowish-green above the knees in breeding birds. Juveniles have all black legs.

Conspicuous features
The biggest of the white-coloured egrets ✐ Black line from gape extends beyond the eye ✐ Strangely kinked neck near middle vertebrae ✐ Yellow bill and black legs in non-breeding birds ✐ Black bill in breeding birds ✐ White back plumes in breeding birds.

Call Grating croak noises when in flight and at the nest.

Nest A platforms of twigs in trees. Up to four bluish-green eggs are laid.

Target localities Miranda, Firth of Thames ✐ Matata Lagoons, Bay of Plenty ✐ Okarito breeding colony, Westland.

Mute Swan

Conspicuous colour White.
Habitat Freshwater lakes and ponds.
Range Scattered throughout New Zealand in selected areas.
Size 1500 mm (House Sparrow 145 mm, Feral Goose 800 mm).
Abundance Uncommon.
Status Introduced.

Family Anatidae

Species *Cygnus olor*

Common names Mute Swan, White Swan

Description – mature
Face Facial skin in front of eye is black.
Upperparts and underparts White.
Bill Orange, with black base and a black knob on the forehead.
Legs and feet Black.

Description – immature
Upper and underparts Grey.

Conspicuous features
Orange bill with black knob at base ✎ The long S-curved neck separates it from a Feral Goose.

Conspicuous characteristic
Leisurely character when swimming.

Call Usually silent but it occasionally trumpets and hisses.

Nest Mound of grass and reeds floating among rushes. Up to seven white eggs are laid.

Target localities Virginia Lake, Wanganui ✎ Lakeside, Lake Ellesmere, near Leeston.

Australasian Gannet (mature)

Conspicuous colours White, yellow and black.
Habitat Coastal and sea.
Range Around New Zealand.
Size 890 mm (House Sparrow 145 mm, Red-billed Gull 370 mm).
Abundance Common.
Status Native.

Family Sulidae **Species** *Morus serrator*

Common name Australasian Gannet, Takapu

Description Forehead White, with black markings in front of eyes. Head Yellow. Upperparts and underparts White. Wings Primary feathers black. Remainder white. Tail Central tail feathers black. Remainder white. Bill and bare skin of face Bluish-grey. Black line around gape. Feet and legs Legs greyish, with feet striped with yellow.

Description – immature Upperparts Brown. Underparts Brown with varying amounts of white until year four when they resemble adults.

Conspicuous feature Yellow head on white body is noticeable.

Conspicuous characteristics Birds feed by diving on to fish from considerable heights with wings folded back. If they catch a fish they usually bob to the surface and eat it before becoming airborne once again. Otherwise they take to the air promptly. Birds are regularly seen flying parallel to the coastline with heads down.

Note The Australasian Gannet (*M. serrator*) is a close relative of the Northern Gannet (*M. bassana*), which is found along both sides of the north Atlantic Sea almost to the Arctic circle. It is also a very close relative of the Cape Gannet (*M. capensis*), which is found around the southern coasts of Africa.

Call Excited high-pitched chatter heard at the breeding colony.

Mainland breeding colonies Muriwai Beach north west of Auckland. Cape Kidnappers south east of Napier. Farewell Spit in the north of the South Island.

Breeding months July to February.

Breeding Birds nest in close proximity to each other with nest spacings of about one metre. One pale blue-green egg is laid in an open nest made of seaweed. The egg laying season is wide, from mid-September to mid-December.

Range worldwide Ranges from New Zealand westward and around southern Australia and up the Queensland coast. Juvenile New Zealand birds migrate to northern Queensland where they spend between three years to seven years before returning to New Zealand to breed.

Target localities Muriwai Gannet Colony north west of Auckland. Cape Kidnappers south east of Napier, Hawkes Bay. Farewell Spit, Nelson.

Southern Giant Petrel (white phase)

Conspicuous colour White.
Habitat Sea.
New Zealand range Around New Zealand for most of the year.
Size 900 mm (House Sparrow 145 mm, Black-backed Gull 600 mm).
Abundance Common.
Status Native.

Family Procellariidae **Species** *Macronectes giganteus*

Common names Giant Petrel, Nelly, Pungurunguru

Phases Comes in a light phase and a dark phase.

Description – mature (white phase) Upper and underparts White with occasional specks of brown. Bill Yellow with a distinctive light-brown tip. Has a prominent nasal tube. Feet and legs Dark grey.

Description – mature (dark phase) Head and neck Light grey. Upper and underparts Brown.

Description – immature Upper and underparts Black.

Conspicuous features White colouring ✐ Yellow bill with light-brown tip ✐ Heavy nasal tube.

Conspicuous characteristics Flight is often straight and direct ✐ It often glides and wheels behind ships ✐ Can glide motionless for some distance.

Breeding islands close to New Zealand Macquarie Island.

Breeding months September to April.

Range worldwide Around southern hemisphere waters.

Target localities Waters off Kaikoura ✐ Waters south from Dunedin.

Yellowhead

Conspicuous colour Yellow.
Habitat Forests.
Range South Island only, mainly confined to Fiordland and Mount Aspiring National Parks. Also in Catlins and Blue Mountains forests in Southland, and the Hawdon River Valley in the Arthurs Pass National Park.
Size 150 mm (House Sparrow 145 mm).
Abundance Rare.
Status Endemic.

Family Pachycephalidae

Species
Mohoua ochrocephala

Common name
Yellowhead, Mohua

Description
Head and underparts Yellow.
Upperparts Olive-brown.
Bill and feet Black.
Eye Black.

Conspicuous features
Yellow head and the black eye and bill ✎ Bright yellow underparts.

Conspicuous characteristics
Feeds in middle to high canopy ✎ Movements are urgent and often noisy ✎ Birds make buzzing notes ✎ If birds are disturbed by a Long-tailed Cuckoo they become very agitated ✎ Will hang upside down when feeding.

Call The song is sweet and gentle and less canary-like than the Whitehead. It is louder than the Brown Creeper. Because it shares a habitat with the Brown Creeper their songs, although different, can sometimes be confused.

Nest It nests in holes of old trees, usually above four metres, making a structure of twigs bound with cobwebs. Up to four pink, lightly speckled with brown, eggs are laid.

Target localities Lake Gunn, Eglinton Valley near Te Anau ✎ Hawdon River Valley, Arthurs Pass.

Yellowhammer

Conspicuous colour Yellow.
Habitat Parks and open country.
Range Throughout New Zealand.
Size 160 mm (House Sparrow 145 mm).
Abundance Common.
Status Introduced.

Family Emberizidae Species *Emberiza citrinella*
Common name Yellowhammer

Description – male bird Head and throat Yellow, with some widely spaced brown markings. Wings Brown, streaked with dark brown. Breast Yellow, washed with brown and streaked with dark brown. Underparts Yellow. Rump Rufous. Tail Brown, with black outer feathers edged with white.

Description – female Upperparts and underparts Similar to male birds but paler and with more brown streaking.

Conspicuous features Yellow plumage ✎ Tail is slightly forked ✎ White outer tail feathers.

Conspicuous characteristics Flocks in large numbers in winter on to pasture where they feed on grass seeds and in particular Poa species ✎ Can be found looking for discarded grass seed in areas where farmers have recently fed out hay.

Call The song, 'chitty chitty chitty chitty swee', made by the male bird, is commonly translated as 'little bit of bread and no cheese'. It is usually delivered from a post or a low shrub. New Zealand birds seem to omit the 'cheese' note. Singing period starts in September and ends in February.

Nest A cup of grass, lined with wool, feathers or moss and usually close to the ground in low growing vegetation or tall grass. Up to four white, scribbled over with brown, eggs are laid. It is a late nester starting in October.

Target localities Common in open country.

Cirl Bunting (male)

Conspicuous colour Yellow.
Habitat Parks and open country.
Range East coast of the South Island from Marlborough to Otago. Occasional North Island sightings.
Size 160 mm (House Sparrow 145 mm).
Abundance Common.
Status Introduced.

Family Emberizidae Species *Emberiza cirlus* Common name Cirl Bunting
Description – male Crown and nape Dark grey, striped with black and yellow.
Face Yellow, with bold black stripe through eye. Chin and throat Black, with yellow

band below. Neck and upperbreast Pale brown. Wings Brown, edged with dark brown. Underparts Pale yellow, lightly streaked with brown. Rump Olive-brown. Tail Olive-brown, with black outer feathers, edged white.

Description – female Upperparts and underparts Dull yellow, streaked heavily with dark brown on the crown and light brown elsewhere. Rump Olive-brown.

Conspicuous features Yellowhammer size ✎ At first sight similar to a Yellowhammer ✎ Overall looks like a dark Yellowhammer.

Conspicuous characteristics Feeds on the ground, especially among fed out hay, with Yellowhammers ✎ Hops while feeding on ground ✎ Often crouches close to ground like a Yellowhammer ✎ Twitters in flight when in family parties.

Call Song, made by the male bird, is a thin, metallic high-pitched rattle, 'tirree tiree tirree tiree' delivered from a conspicuous perch. It is not unlike the Yellowhammer's song. Song period is from October to March, this bird still singing when all the other introduced songsters have stopped. Birds also call when in flight and have a variety of zit type communication calls.

Nest A cup of dried grass and moss in thick vegetation at about two metres above the ground, higher than the Yellowhammer. Up to four bluish green, streaked with black eggs are laid.

Target localities Nelson district ✎ Motueka district ✎ Timaru district and racecourse.

Differences between female Cirl Buntings and female Yellowhammers

- Cirl Buntings have olive-brown coloured rumps.
- Yellowhammers have rufous coloured rumps.
- Cirl Buntings are darker with less yellow.

Differences between male Cirl Buntings and male Yellowhammers

- Cirl Buntings have a greyish crown.
- Yellowhammers have yellow crowns.
- Cirl Buntings have a yellow face separated by a bold black eye-stripe.
- Cirl Buntings have a black chin.

Glossary

Circumpolar found breeding naturally in New Zealand but also in colder latitudes.

Endemic found breeding naturally only in the New Zealand region.

Exotic introduced from another country.

Indigenous belonging naturally and breeding in a more particular area within a country.

Introduced species that have been brought here by people.

Native self-introduced and found breeding naturally in New Zealand but also in other countries.

Stragglers species of birds that regularly visit in very small numbers.

Vagrants species that turn up on very rare occasions.

Photograph sources

Robin Bush
Pages 23, 24, 29 upper and lower, 30, 31, 33, 34, 36, 37, 42, 44 left and right, 46, 49, 50, 53, 55, 57 upper, 62, 65 lower, 66 upper, 67, 69, 70 left and right, 73, 75, 76 right, 77, 78, 81, 82, 83, 85, 86, 87, 89 inset, 91, 94, 96, 97, 100, 103, 106, 107, 116, 118, 119, 121, 125, 136, 137, 142, 143, 145, 146 upper and lower, 148, 151, 154, 155, 156, 159, 160, 161, 162, 166, 167, 168, 170 left, 171, 172, 173, 174 left and right, 175, 176, 178, 180, 181, 183, 185, 186, 188, 189, 190, 191, 195, 198, 199, 202, 204, 205, 206, 207, 208, 209, 211, 216, 222, 224, 225, 226, 227, 230 lower, 231 left, 233, 234, 237, 238, 240 upper and lower, 242, 243, 244, 245, 248, 249, 250, 251, 252, 253, 255, 256, 257, 258, 266, 267 upper and lower, 268, 269, 270, 271, 273, 274.

Simon Fordham
Pages 22, 27, 28, 35, 38 upper, 45 upper and lower, 51, 52, 56, 80, 88, 90, 93, 95, 101, 110, 112, 114, 115, 122, 123, 124, 128, 129, 130, 132, 134, 144, 147, 149, 150, 152, 157, 164, 187, 194, 201, 210, 215 upper, 219, 229, 230 upper, 239 upper, 247, 259, 260, 261.

Ian Southey
Pages 21, 32, 47 right, 57 upper, 61, 98 upper and lower, 99, 108, 109, 113, 120, 138, 163, 170 right, 184, 217, 218, 220, 236, 241.

Ray Wilson
Pages 25, 26, 38 lower, 39, 40 upper and lower, 43 upper, 54, 59 upper, lower and inset, 63, 68, 72, 76 left, 105, 131 main, 135, 169, 179 main, 200, 215 lower, 239 lower, 263, 264, 275.

Stuart Chambers
Pages 58, 60, 65 upper, 66 lower, 89 main, 111, 131 insets, 140, 165, 212, 213, 231 right, 254.

Dennis Buurman
Pages 74, 158, 192, 232 inset, 272.

Peter La Tourrette
Pages 182, 193, 221.

Ian Montgomery
Page 153.

Bill Jolly
Pages 262, 265.

Ian Tew
Page 126.

Mike Fuller
Pages 57 lower, 102, 157.

Department of Conservation
Pages 41, 43 lower, 47 left, 71, 92, 127, 133 left and right, 139, 141, 179 inset, 196, 197, 203, 232 main, 246.

Note:
In the compilation of this book we have handled many pictures. We have done our best to link them all with the correct photographers but, if a mistake has been made, please accept our apologies and notify the publisher. It will be corrected in any future edition.

General index

Index of Latin names